AMERICAN WAR PROPAGANDA LEAFLETS

1990-2022

AMERICAN WAR PROPAGANDA LEAFLETS

1990-2022

Printed in the United States of America

"Capture their minds, and their hearts and souls will follow."

— An old-school PSYOP motto.

Publisher's Note

In 2022, Khajistan began acquiring hundreds of printed items from collectors in Pakistan, Afghanistan, and the U.S., originally dropped by the U.S. military and CIA in Afghanistan, Iraq, and Libya over the past three decades. Now housed at the Kislak Center at the University of Pennsylvania, this collection provides a raw look at America's wartime messaging.

This book offers a curated selection of these propaganda pieces for Western and westernized audiences. It includes some leaflets never distributed, used instead as prototypes or for morale. Double-sided leaflets are treated as separate entries, and we've added other ephemera like "safe passage cards." A glossary and detailed catalog of the leaflets' imagery and objectives are included at the end.

Dedication

To the people of Afghanistan, Iraq, and Libya affected by the destructive force of the American Military-Industrial Complex.

To Herbert A. Friedman, for his lifelong dedication to documenting psychological propaganda.

CONTENTS

THE GULF WAR

Operation Desert Shield

(August 7, 1990 – January 16, 1991)

Leaflet 1

Warning!

This is just the beginning!

So, this bomb is real.

We want to inform them that we want to protect the innocent people but Saddam is leading them in the path of death and inevitable death.

Indeed!

Saddam is the reason.

Yes, the Joint Forces are fully capable of attack at any time or place.

Warning!

Arabic

Leaflet 2

Warning!
This site will be bombed soon.
Leave your equipment and save yourself.
warning!

Arabic

Leaflet 3

it is too late!

Arabic

Leaflet 4

This site is being bombed!

Arabic

Leaflet 5

Joint Forces

Arabic

Leaflet 6

Saddam is the only reason to bomb Iraq!

Arabic

Leaflet 7

ان أعمال صدام حسين هي التي جعلت العالم يحارب العراق، وتبذل
القوات المشتركة الجوية أقصى جهودها لتتجنب اصابة المدنيين الأبرياء،
إلا أن صدام وضع بعض المواطنين العراقيين في المراكز العسكرية ليموتوا
بدلا من زبانيته، لهذا فهو مستعد للتضحية بكم جميعاً وبالاماكن المقدسة
وبتاريخ العراق مقابل بقائه.

It is the actions of Saddam Hussein which have forced the world to war with Iraq. The Joint Air Forces are making a great effort to avoid injuring innocent civilians. If Saddam puts Iraqi civilians in military headquarters to die instead of his thugs, then by this he shows he is ready to sacrifice all of you, the holy places, and the history of Iraq for his survival.

Arabic

Leaflet 8

Leaflet 9

Saddam's first line of defense consisted of innocent civilians.

Arabic

Leaflet 10

My brother, the Iraqi soldier have you thought about the strength of the Joint Forces?

Arabic

Leaflet 11

You are isolated, stop resisting.

Arabic

Leaflet 12

Leaflet 13

Oh, my child, when will you return?

Arabic

Leaflet 14

My brother, the Iraqi soldier, it pains us deeply that you would return to Iraq either killed or disabled

Arabic

Leaflet 15

Leaflet 16

Leaflet 17

Leaflet 18

أنتم عزل !!

فقد إنقطعت خطوط امداداتكم ولن يصل اليكم ابدا
التموين والتعزيز فلا يهم صدام مصيركم وهو
ترككم للقاء مصيركم. ألقوا اسلحتكم،
وانضموا مع اخوانكم العرب في حب وسلام !!!

You are defenseless!!

Your supply lines have been cut off, and supplies and reinforcements will never reach you. Saddam does not care about your fate, and he left you to meet your fate. Drop your weapons. Join with your Arab brothers in love and peace!!!!

Arabic

Leaflet 19

Leaflet 20

Leaflet 21

Staying here means death.

Arabic

Leaflet 22

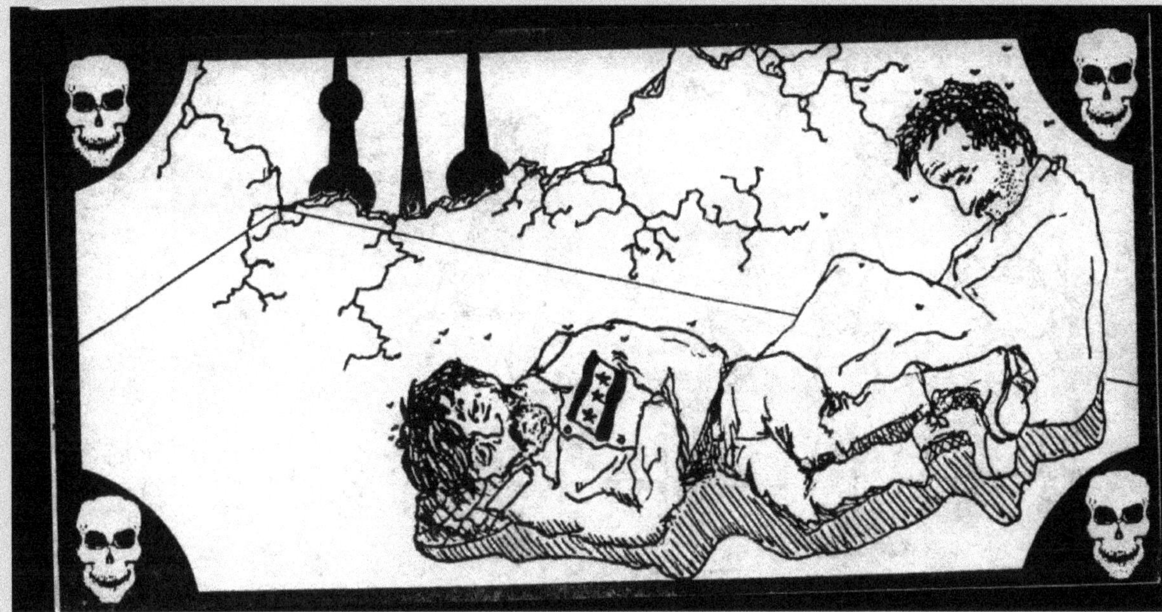

Leaflet 23

Flee to your homes.

Arabic

Leaflet 24

The 7th Corps of the Multi-National Forces is heading towards you soon.
Your colleagues in front of you have surrendered or been eliminated.
Your turn will come soon.

Arabic

Leaflet 25

The cold face of death and war.

Arabic

Leaflet 26

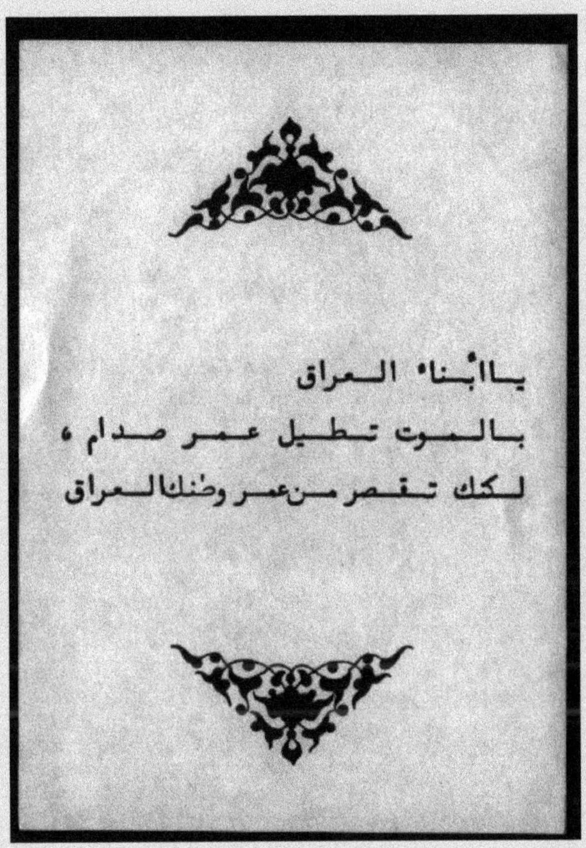

يا أبناء العراق
بالموت تطيل عمر صدام ،
لكنك تقصر من عمر وطنك العراق

O sons of Iraq

By dying, you prolong Saddam's life, but you shorten the life of your homeland, Iraq.

Arabic

Leaflet 27

I crossed the river for you, and I placed you on my head (a phrase implying great respect or priority). With every attack, I feel death, and my breath weakens.

Shatt Al-Arab

Arabic

Leaflet 28

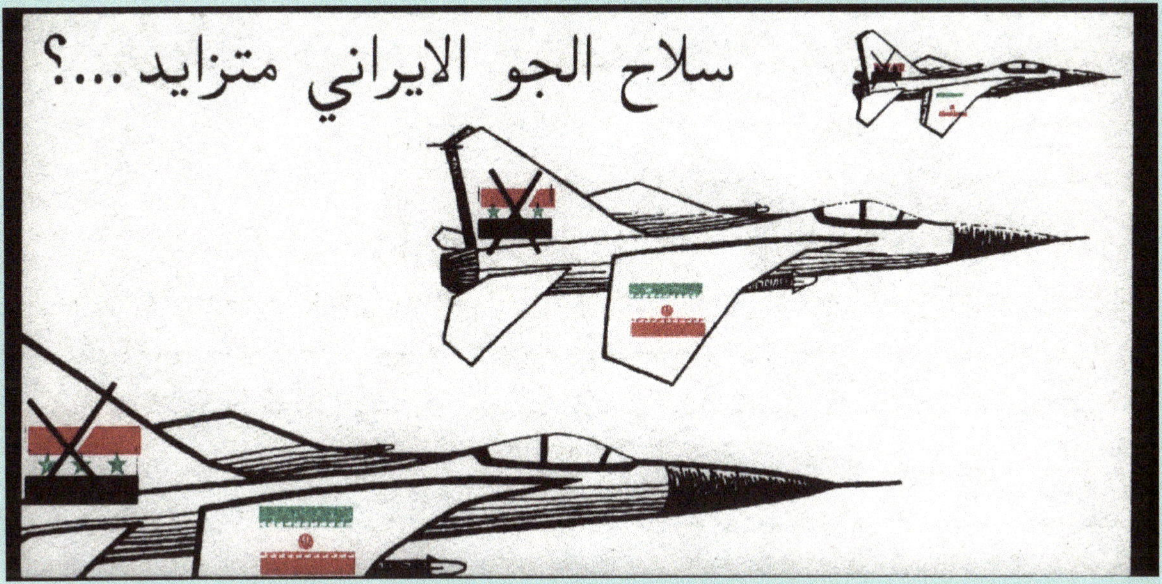

Iranian air force is growing...?

Arabic

Leaflet 29

لقد بادر صدام في اعادة شط العرب لايران والان.....
يقدم لهم سلاحكم الجوي
لكن هناك خطوة افضل بامكانكم اتخاذها وهي الذهاب الى المملكة العربية السعودية.
اعلنوا بواسطة اجهزة الارسال عن رغبتكم بالالتجاء الى اخوانكم العرب. اقترب بطائرتك
بشكل انفرادي، انزل عجلات الهبوط وسلط كافة انوار الطائرة، عندها اطرح جميع
الذخائر الموجودة على متنها واقفل اجهزة تصويب الاهداف ثم تابع التحليق
بسرعة لا تتجاوز ٢٥٠ الى ٣٠٠ عقدة في الساعة.
انك ستحصل على معاملة انسانية، وعند انتهاء الازمة سيسمح
لك بالعودة الى وطنك لاعادة بنائه من جديد.

Saddam has given back the Shatt al Arab to Iran, and now he gives them your Air Force. There is something better you can do. Go to Saudi Arabia. Call us on your radio and announce that you wish to join your Arab brothers. Approach us alone in your aircraft, or in a single file. Lower your landing wheels, turn on your landing lights, disarm your weapons, turn off your fire control systems and continue flying at a speed no greater than 250 to 300 nautical miles. You will receive humane treatment, and after the crisis is over, you will be permitted to return to your homeland and rebuild it.

Arabic

Leaflet 30

If you want to save your life, follow these steps:

Remove the ammunition magazine from your weapon.

Raise your weapon above your head with your left hand.

Place the weapon on the ground gently, with the barrel pointed down.

To ensure your safety, keep your hands raised above your head and move slowly toward the designated safe area, where the flag of peace is displayed.

You are welcome here.

Arabic

Leaflet 31

Please tune your radio to the medium wave frequency of 1134 kHz daily.

Arabic

Leaflet 32

"Listen to Radio Kuwait on FM frequencies 92.5, 97.5, 87.9, and 98.8 MHZ.

Arabic

Leaflet 33

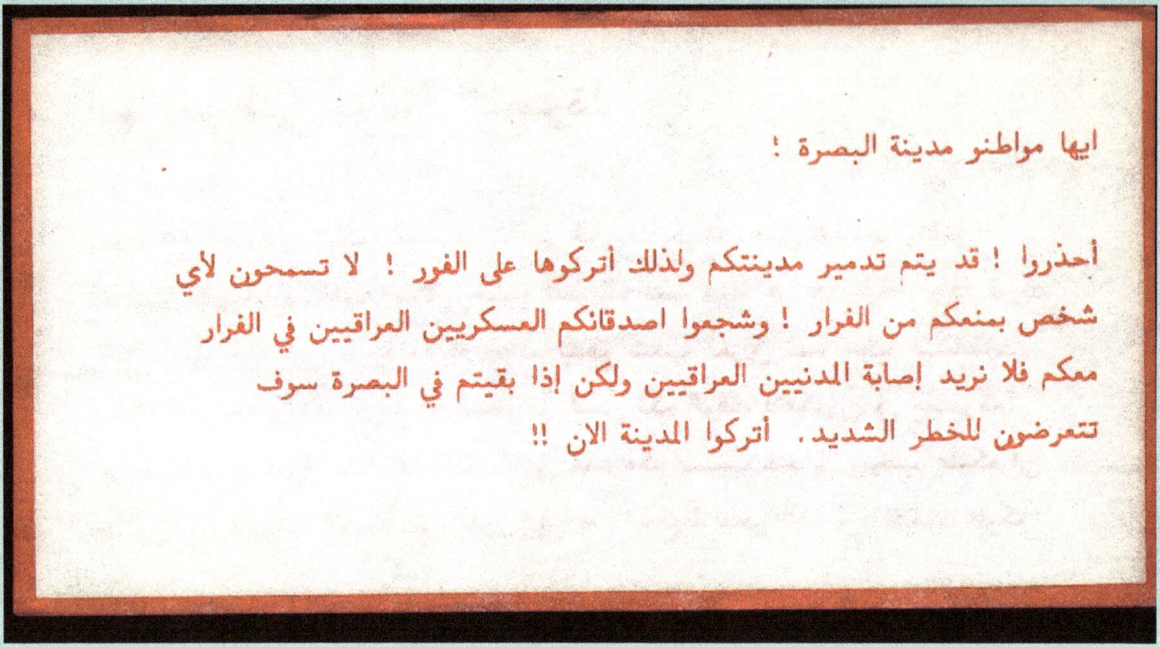

ايها مواطنو مدينة البصرة !

أحذروا ! قد يتم تدمير مدينتكم ولذلك أتركوها على الفور ! لا تسمحون لأي
شخص بمنعكم من الفرار ! وشجعوا اصدقائكم العسكريين العراقيين في الفرار
معكم فلا نريد إصابة المدنيين العراقيين ولكن إذا بقيتم في البصرة سوف
تتعرضون للخطر الشديد. أتركوا المدينة الان !!

O citizens of Basra!

Be careful. Your city could be destroyed. Leave at once! Do not allow anyone to stop your escape from the city! Encourage your military friends to flee with you. If you stay in Basra, you will expose yourself to great danger. Leave the city at once!!

Arabic

Leaflet 34

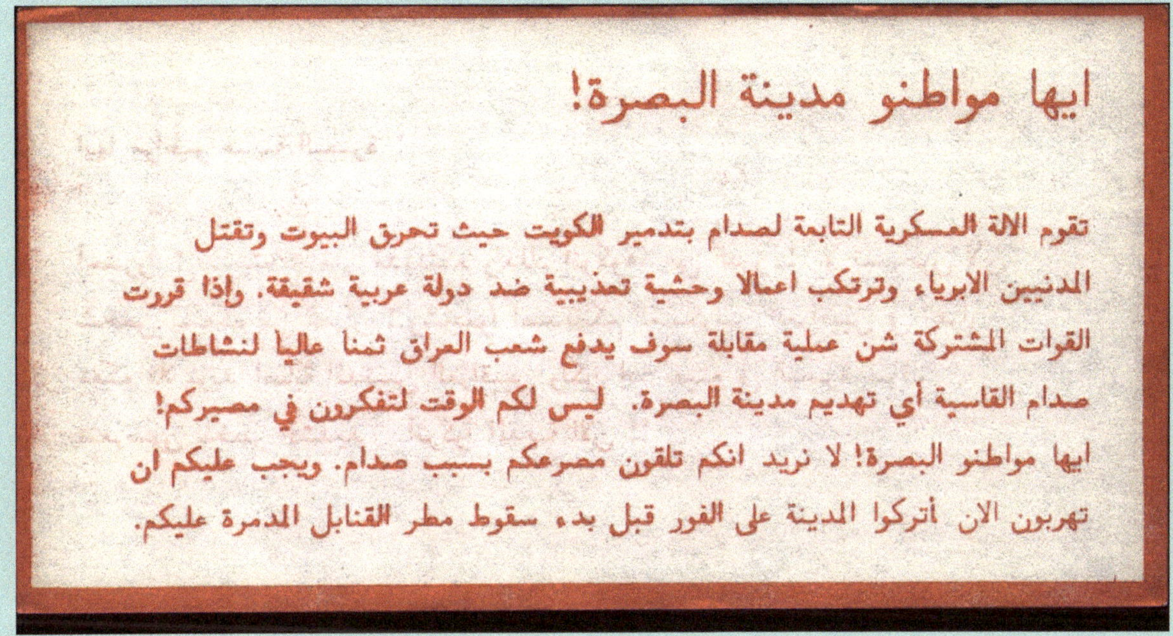

Dear citizens of the city of Basra

Saddam's military machine is destroying Kuwait, burning homes, killing innocent civilians, and committing brutal acts of torture against a sister Arab country. If the joint forces decide to launch a corresponding operation, the people of Iraq will pay a high price for Saddam's cruel activities, namely the demolition of the city of Basra. You do not have time to think about your fate, citizens of Basra. We do not want you to die because of Saddam. You must flee now, leave the city immediately before the rain of devastating bombs begins to fall on you.

Arabic

Leaflet 35

STAY IN YOUR HOMES!! DO NOT APPROACH THE COALITION FORCES!! HOSTILE ACTION IS IMMINENT IN THIS AREA!! STAY IN YOUR HOMES AND YOU WILL BE SAFE!! TO APPROACH THE COALITION FORCES IS TO PLACE YOUR LIFE IN DANGER!! THE COALITION FORCES WISH YOU NO HARM!! STAY IN YOUR HOMES!!

English

Leaflet 36

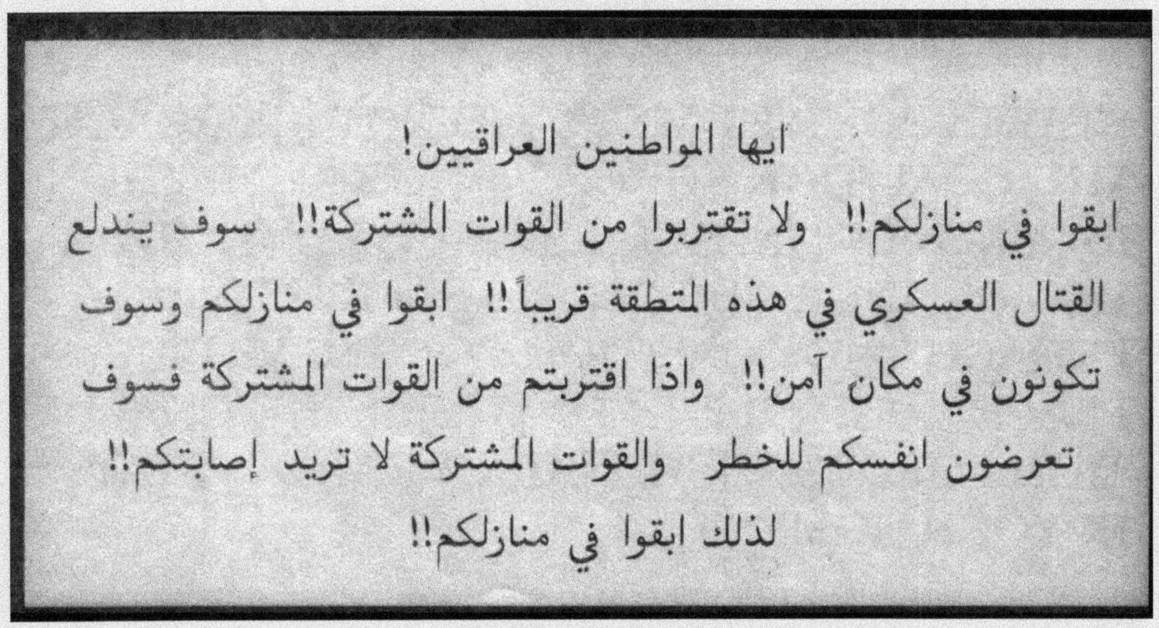

ايها المواطنين العراقيين!
ابقوا في منازلكم!! ولا تقتربوا من القوات المشتركة!! سوف يندلع
القتال العسكري في هذه المنطقة قريباً!! ابقوا في منازلكم وسوف
تكونون في مكان آمن!! واذا اقتربتم من القوات المشتركة فسوف
تعرضون انفسكم للخطر والقوات المشتركة لا تريد إصابتكم!!
لذلك ابقوا في منازلكم!!

Dear Iraqi citizens, stay at home!! Do not approach the joint forces!! Military fighting will break out in this area soon!! Stay home and you will be safe!!! If you approach the joint forces, you will expose yourselves to danger, and the joint forces do not want to harm you!!!

So, stay at home!!

Arabic

Leaflet 37

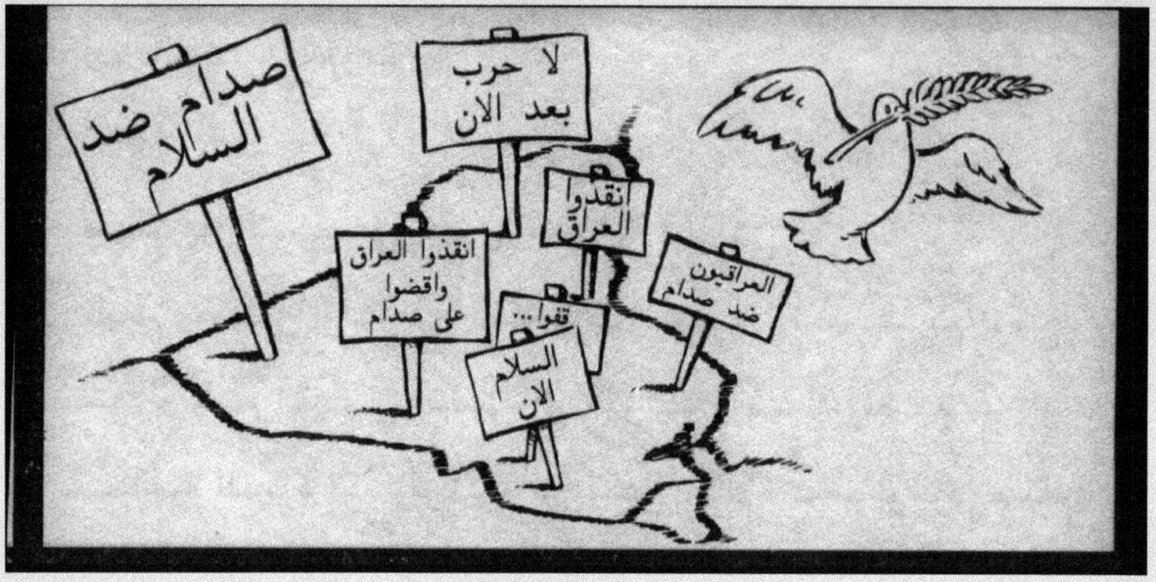

Saddam is against peace.

Help Iraq, stop Saddam.

No more war.

Peace now.

Help Iraq,

Iraq unites against Saddam.

Confront...

Arabic

Leaflet 38

ايها المواطنون العراقيون

السلام عليكم

صدام هو المتسبب في الحرب ونتائجها لكم هي – ارامل

وايتام ومعوقون ومشوهون ولكم الخيار انتم لا احد غيركم بوضع

حد له وانتم القادرون انضموا الى اخوانكم واظهروا وفضكم

لسياسته المدمرة لكم ولوطنكم لن يكين هناك سلام بوجود صدام.

Dear Iraqi citizens

Hello

Saddam is the cause of the war and its sorrows. He must be stopped. Join with your brothers and demonstrate the rejection of Saddam's brutal policies. There will be no peace with Saddam.

Arabic

Leaflet 39

My Iraqi brother, what we want is peace.

Arabic

Leaflet 40

Leaflet 41

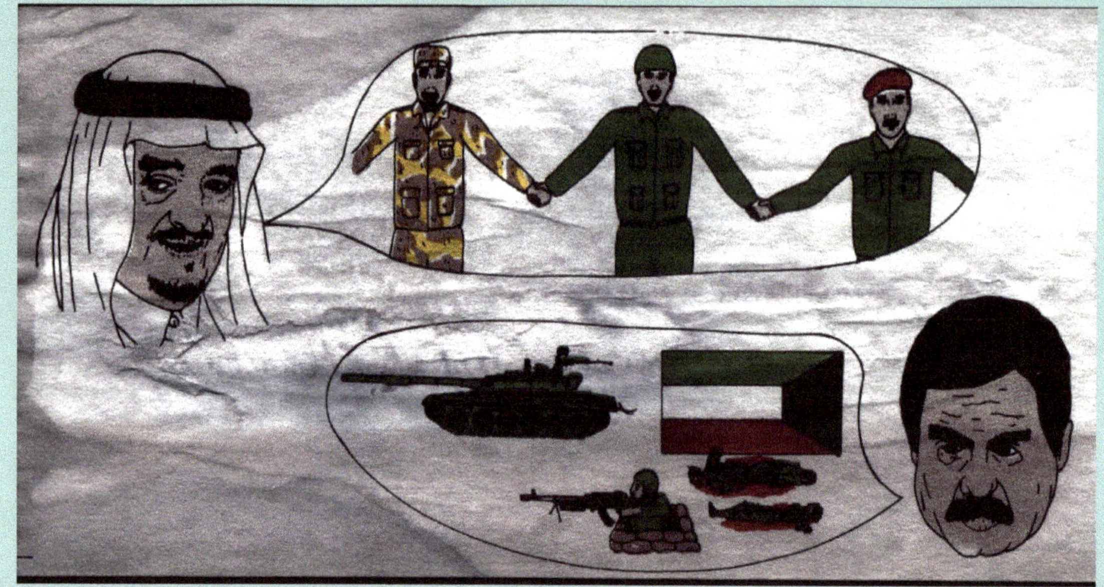

Leaflet 42

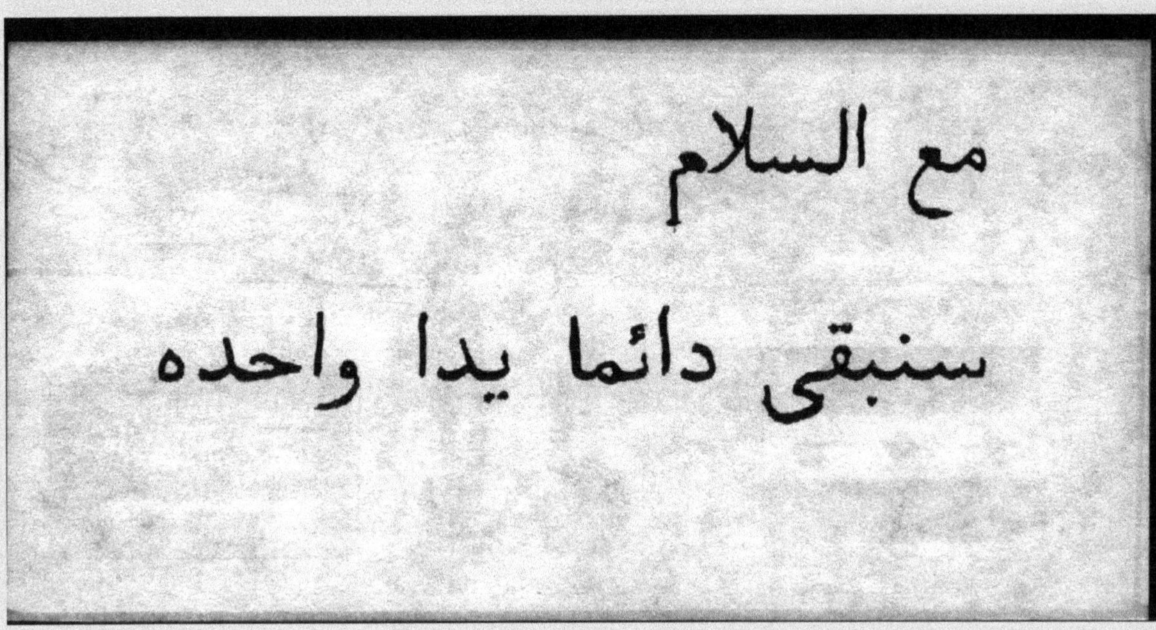
مع السلام

سنبقى دائما يدا واحده

In peace we will always remain united.

Arabic

Leaflet 43

Warning.

Don't touch anything suspicious that encourages your curiosity such as radios, cans, handbags, and mechanical machines. They may explode when touched. Report all strange objects to the authorities or the Joint Forces immediately.

Arabic

Leaflet 44

قيادة القوات المشتركة ومسرح العمليات

بطاقة دعوة

في ضيافة قيادة القوات المشتركة ومسرح العمليات. انت مدعو للانضمام للقوات
المشتركة مع تمتعك الكامل بواجب الضيافة العربية والأمن والسلامة والرعاية
الطبية . . والعودة للأهل بمجرد الانتهاء من الوضع الذي وضعنا فيه صدام.

أخي الجندي العراقي .

هذه دعوة مفتوحة لك ولأخوانك المقاتلين نرجو أن تلبيها عندما تسمح لك الفرصة.

قائد
القوات المشتركة ومسرح العمليات

Joint Forces Command and Theater of Operations

Invitation Card

Under the hospitality of the Joint Forces Command of the coalition, you are invited to join the joint forces with full safety, security, and medical care... and return to your family safely. From today onwards, Saddam has put us in this situation.

My Iraqi soldier brother, This is an open invitation for you and your fellow fighters.

This is your opportunity.

Commander of the Joint Forces and Theater of Operations

Arabic

Leaflet 45

Leaflet 46

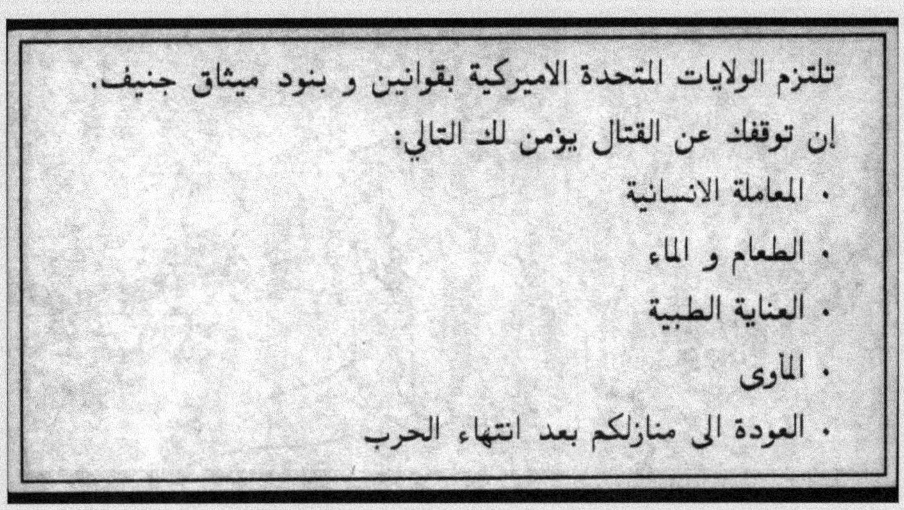

تلتزم الولايات المتحدة الاميركية بقوانين و بنود ميثاق جنيف.

إن توقفك عن القتال يزمن لك التالي:

- المعاملة الانسانية
- الطعام و الماء
- العناية الطبية
- المأوى
- العودة الى منازلكم بعد انتهاء الحرب

The United States of America is committed to the laws and provisions of the Geneva Convention.

If you stop fighting, you will gain the following:

- Humane treatment
- Food and water
- Medical care
- Shelter
- Return to your homes after the war ends.

Arabic

Leaflet 47

Operation Desert Storm

(January 17, 1991 – February 28, 1991)

Leaflet 48

Desert Storm (Sahara Whirlwinds) is coming to your area. Flee immediately.

Arabic

Leaflet 49

Staying here means death.

Arabic

Leaflet 50

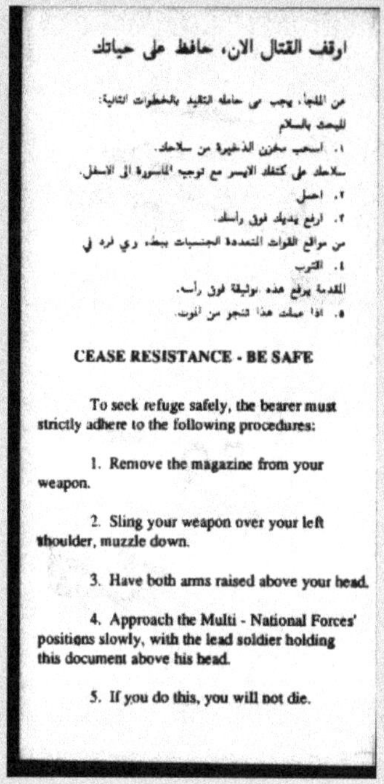

اوقف القتال الان، حافظ على حياتك

من الملجأ، يجب على حامله التقيد بالخطوات التالية:
للبحث بالسلام
١. اسحب مخزن الذخيرة من سلاحك.
سلاحك على كتفك الايسر مع توجيه الماسورة الى الاسفل.
٢. احمل
٣. ارفع يديك فوق رأسك.
من مواقع القوات المتعددة الجنسيات ببطء وي فرد في
٤. اقترب
المقدمة يرفع هذه الوثيقة فوق رأسه.
٥. اذا عملت هذا تنجو من الموت.

CEASE RESISTANCE - BE SAFE

To seek refuge safely, the bearer must strictly adhere to the following procedures:

1. Remove the magazine from your weapon.

2. Sling your weapon over your left shoulder, muzzle down.

3. Have both arms raised above your head.

4. Approach the Multi - National Forces' positions slowly, with the lead soldier holding this document above his head.

5. If you do this, you will not die.

CEASE RESISTANCE - BE SAFE

To seek refuge safely, the bearer must strictly adhere to the following procedures:

Remove the magazine from your weapon.

Sling your weapon over your left shoulder, muzzle down.

Have both arms raised above your head.

Approach the Multi-National Forces' positions slowly, with the lead soldier holding this document above his head.

If you do this, you will not die.

Arabic, English

Leaflet 51

ايها المواطنون العراقيون
قدعرضتكم قوات صدام العسكرية الى الخطرالشديد .
ان القوات المتحالفة تتقدم نحوكم بسرعة هائلة و سنبدأ
بقصف هذا الموقع قريباً و لا نريد اصابة المدنيين الابرياء .
اخلوا هذه المنطقة واتجهوا الى الشمال ، إن الاحياء السكنية
في مدينة بغداد لن تتعرض للقصف .
اهربوا فوراً !!!

Iraqi citizens.

Saddam's military has placed your lives in danger. The Coalition forces are coming. We will begin striking this area soon. We do not wish to harm innocent civilians. Evacuate this area immediately and head north. Civilian areas in Baghdad will not be targeted.

Flee Immediately!!!!

Arabic

Leaflet 52

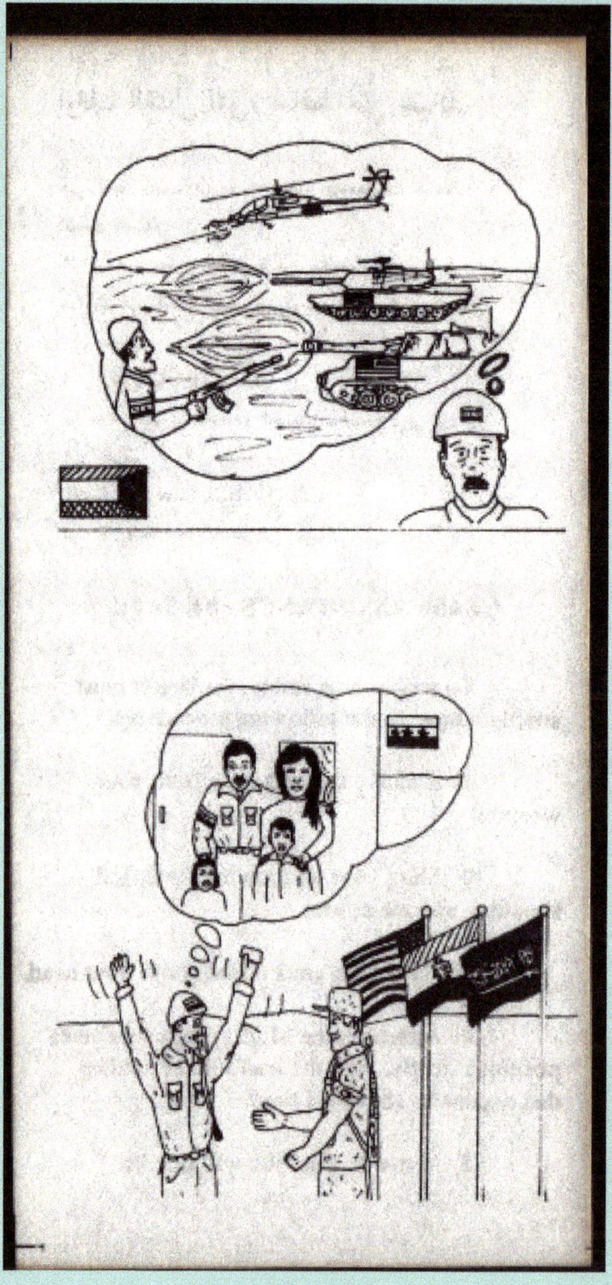

Leaflet 53

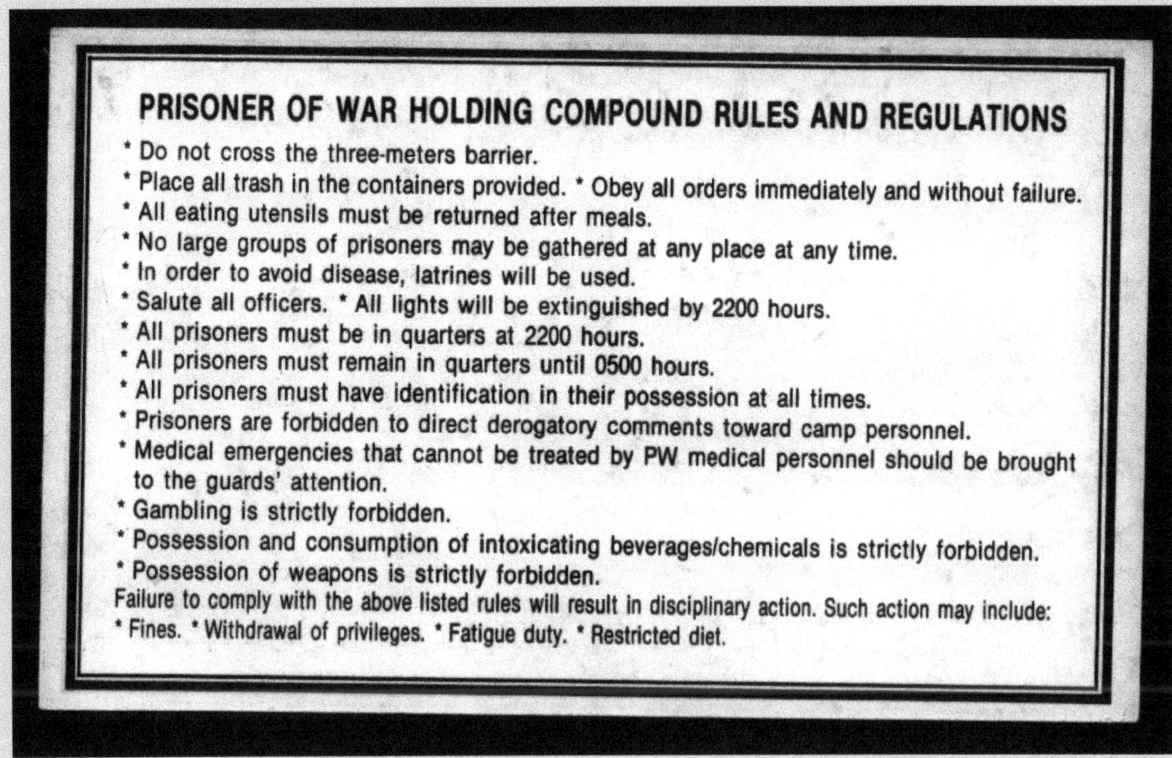

PRISONER OF WAR HOLDING COMPOUND RULES AND REGULATIONS

* Do not cross the three-meters barrier.
* Place all trash in the containers provided. * Obey all orders immediately and without failure.
* All eating utensils must be returned after meals.
* No large groups of prisoners may be gathered at any place at any time.
* In order to avoid disease, latrines will be used.
* Salute all officers. * All lights will be extinguished by 2200 hours.
* All prisoners must be in quarters at 2200 hours.
* All prisoners must remain in quarters until 0500 hours.
* All prisoners must have identification in their possession at all times.
* Prisoners are forbidden to direct derogatory comments toward camp personnel.
* Medical emergencies that cannot be treated by PW medical personnel should be brought to the guards' attention.
* Gambling is strictly forbidden.
* Possession and consumption of intoxicating beverages/chemicals is strictly forbidden.
* Possession of weapons is strictly forbidden.

Failure to comply with the above listed rules will result in disciplinary action. Such action may include:
* Fines. * Withdrawal of privileges. * Fatigue duty. * Restricted diet.

English

Leaflet 54

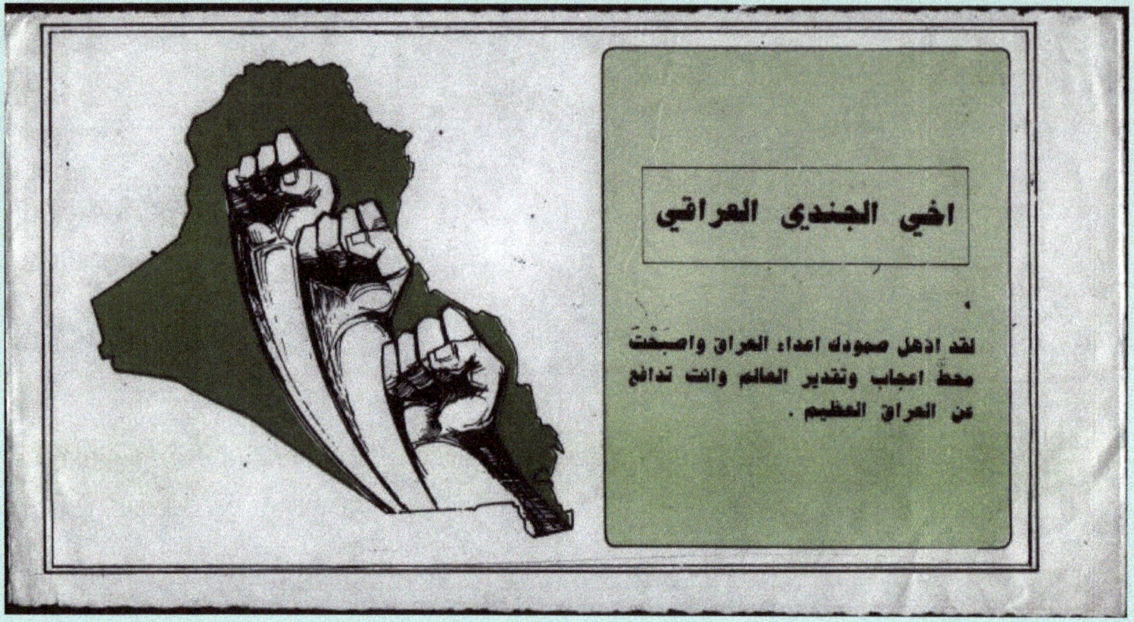

My brother Iraqi soldiers Your resistance distracted the enemies of Iraq You have been placed at the center of attention and appreciation of the whole world while you stood to protect the great Iraq

Arabic

Leaflet 55

My brother, the Arab soldier:

Who do you point your weapon at? ... Remember that the one you are fighting today is the one whose blood mixed with your blood in Palestine, Sinai, and the Golan. How can you compel yourself to remain in the trenches of the American-Zionist alliance against Iraq?

Arabic

Leaflet 56

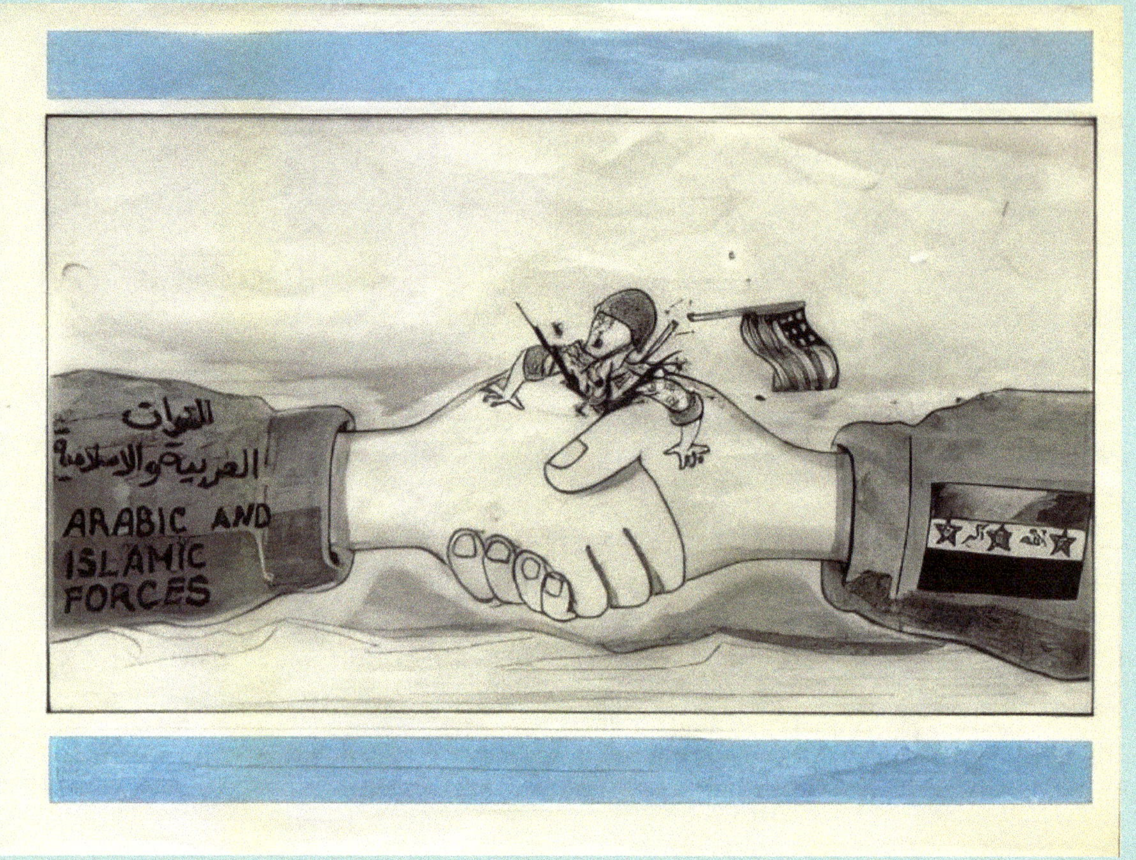

Allah Akbar. God is Great.

Arabic and Islamic Forces.

Arabic and English

Leaflet 57

لوائح وقواعد معسكرات اعتقال اسرى الحرب

- ابتعد عن الحواجز ثلاثة أمتار
- ضع كل المهملات في الأوعية المعدة لذلك ـ يجب إعادة كل أواني الأكل بعد الانتهاء منها
- لا يسمح بتجمعات كبيرة من الأسرى في أي مكان وفي أي وقت .
- تفادياً للأمراض يجب استخدام المراحيض .
- القي التحية العسكرية على جميع الضباط ـ تطفأ جميع الأنوار في الساعة العاشرة مساءً
- يجب تواجد جميع الأسرى في أماكنهم في الساعة العاشرة مساءً
- يجب بقاء جميع الأسرى في أماكنهم حتى الساعة الخامسة صباحاً
- يجب أن تكون بطاقات الهوية في حيازة الأسرى في كل الأوقات
- يجب على جميع الأسرى الامتناع عن توجيه عبارات سيئة لأفراد المعسكر
- الحالات الطبية الطارئة التي لا يمكن علاجها بواسطة الفريق الطبي للأسرى يجب اعلام الحراس عنها
- القمار (الميسر) ممنوع منعاً باتاً
- حيازة أو شرب المشروبات الروحية/ الكيمائية ممنوع منعاً باتاً ـ حيازة الأسلحة ممنوع منعاً باتاً

إن عدم مراعاة التعليمات الموضحة عاليه سوف يترتب عليها إجراءات نظامية تشمل ما يلي :

• الغرامات • الحرمان من الامتيازات • القيام بأعمال النظافة وغيرها • غذاء محدود (غير كاف)

Arabic

Rules and Regulations for War Prisoner Detention Camps

Stay three meters away from the barriers.

All lights must be turned off at 10:00 PM. Salute all officers. All prisoners must be in their places by 10:00 PM.

Dispose of all trash in the designated containers. All eating utensils must be returned after use.

Large gatherings of prisoners are not allowed at any place or time to prevent disease. Use the toilets.

All prisoners must remain in their places until 5:00 AM. Identification cards must be in the possession of prisoners at all times.

Medical emergencies that cannot be treated by the prisoners' medical team must be reported to the guards.

All prisoners must refrain from directing insults at camp personnel.

Gambling is strictly prohibited.

Possession or consumption of alcoholic or chemical beverages is strictly prohibited. Possession of weapons is strictly prohibited.

Failure to comply with the above instructions will result in disciplinary actions, including:

Fines

Deprivation of privileges

Assignment of cleaning duties and other tasks

Limited (insufficient) food.

Arabic

Leaflet 58

Time isn't on your side... How many days are left? ...

Keep counting... until January 15th!!

Arabic

Leaflet 59

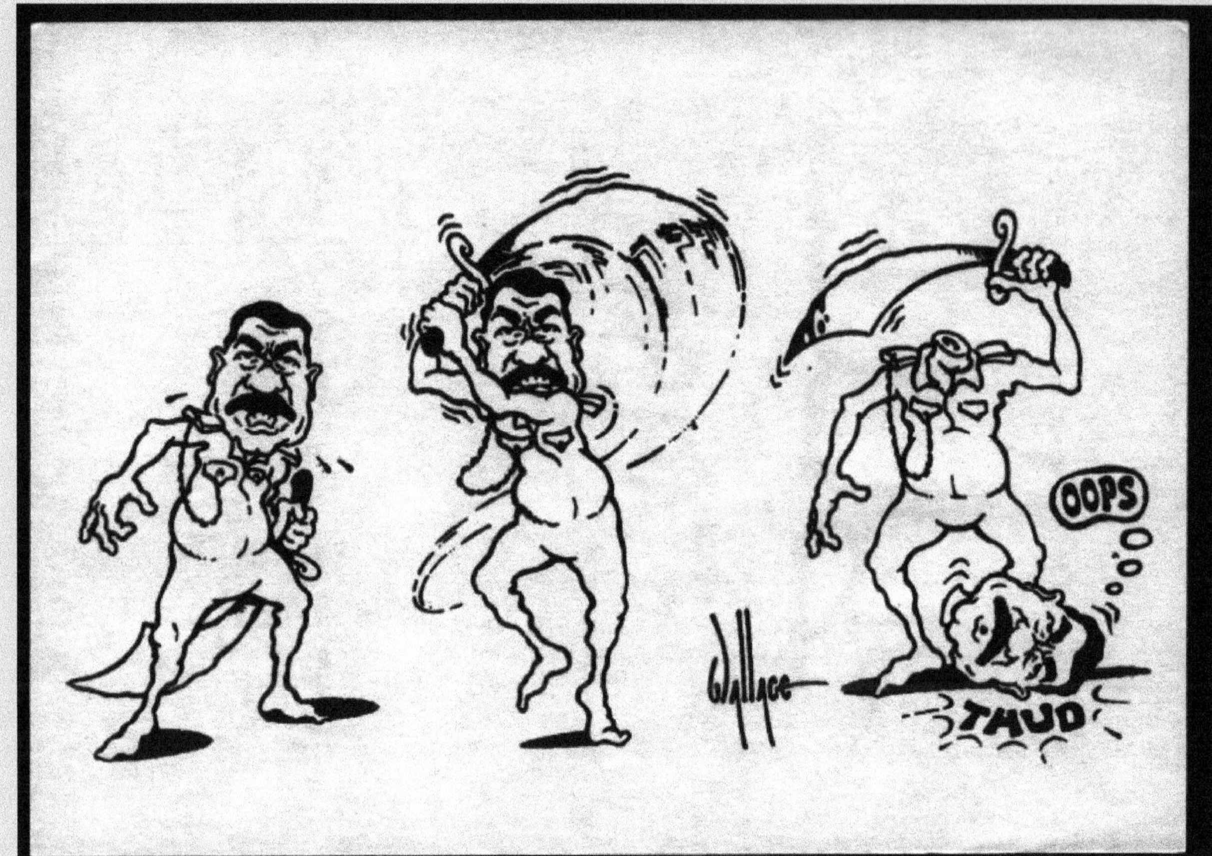

Leaflet 60

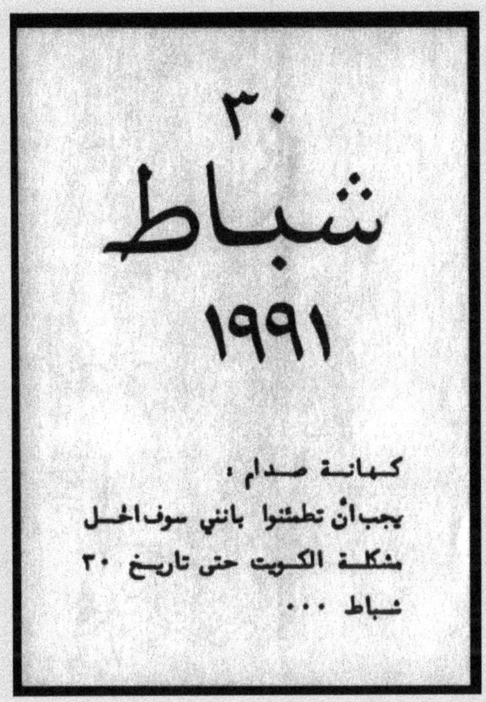

30 February 1991.

Saddam's prediction: Be assured that I will solve the problem of Kuwait by the 30th of February...

Arabic

Leaflet 61

Activities scheduled جدول المواعيد

أ) الاستيقاظ - الساعة ٥ صباحاً
ب) عملية حصر الحضور - الساعة ٥,٣٠ صباحاً
جـ) الإفطار - الساعة ٥,٤٥
د) عملية تسوية وصيانة المعسكر - الساعة ٦,٣٠
هـ) بيان تفاصيل العمل - الساعة ٧,٣٠
و) العيادة عند بوابة المعسكر - الساعة ٧,٣٠
ز) عملية حصر الحضور - الساعة ١١,٤٥ ظهراً
ح) وجبة الغذاء - الساعة ١٢,٠٠ ظهراً
ط) عملية حصر الحضور - الساعة ٤,٤٥ عصراً
ي) وجبة العشاء - الساعة ٥,٠٠ مساءً
ك) كافة الأسرى يكونون في المعسكر - الساعة ٧ مساء (ما عدا الأوقات الخاصة بمتطلبات الحمامات والنظام الصحي الأخرى)
ل) إطفاء الأنوار الساعة ١٠ مساءً - يظل أسرى الحرب في مضاجعهم بين الساعة ١٠ مساءً والخامسة صباحاً

Activities scheduled:

A) Waking up - 5 am
B) Attendance counting process - 5:30 am
C) Breakfast - 5:45
D) Camp settlement and maintenance process - 6:30
E) Statement of work details - 7:30
F) The clinic at the camp gate - at 7:30.
G) Attendance counting process - at 11:45 pm
H) Lunch - 12:00 noon
I) Attendance counting process - 4.45 pm
J) Dinner - 5:00 pm
K) All prisoners shall be in camp at 7 p.m. (except during times for bathroom and other sanitary requirements) in the evening. Prisoners of war shall remain in.
L) Turn off the lights at 10 p.m. They sleep between 10 pm and 5 am

Arabic

Leaflet 62

Yesterday, we presented to you proof of the ability and superiority of the Joint Forces. For the second time we are giving you another chance to survive.

Arabic

Leaflet 63

The 48th Infantry Division will be bombed tomorrow.
Leave this position now and save yourself.

Arabic

Leaflet 64

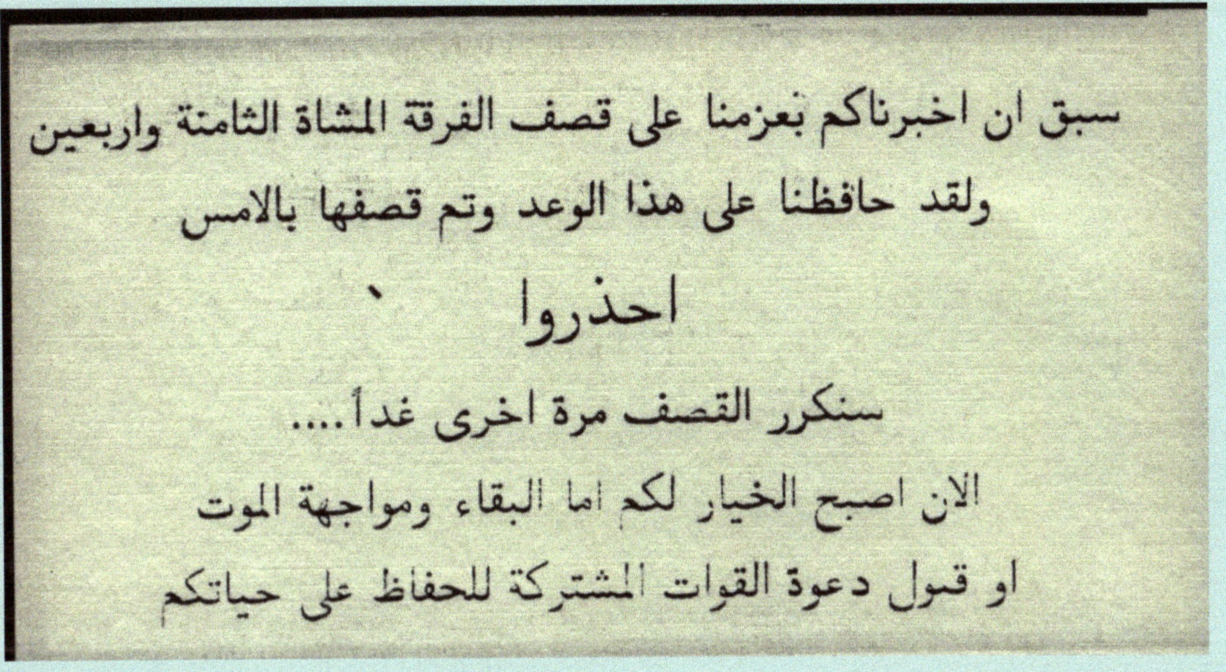

سبق ان اخبرناكم بعزمنا على قصف الفرقة المشاة الثامنة واربعين

ولقد حافظنا على هذا الوعد وتم قصفها بالامس

احذروا

سنكرر القصف مرة اخرى غداً....

الان اصبح الخيار لكم اما البقاء ومواجهة الموت

او قبول دعوة القوات المشتركة للحفاظ على حياتكم

We previously informed you of our intention to bomb the 48th Infantry Division, and we kept this promise, and it was bombed yesterday.

Be careful.

We will repeat the bombing again tomorrow..... Now the choice is yours: either stay and face death or accept the call of the Joint Forces to preserve your lives.

Arabic

Leaflet 65

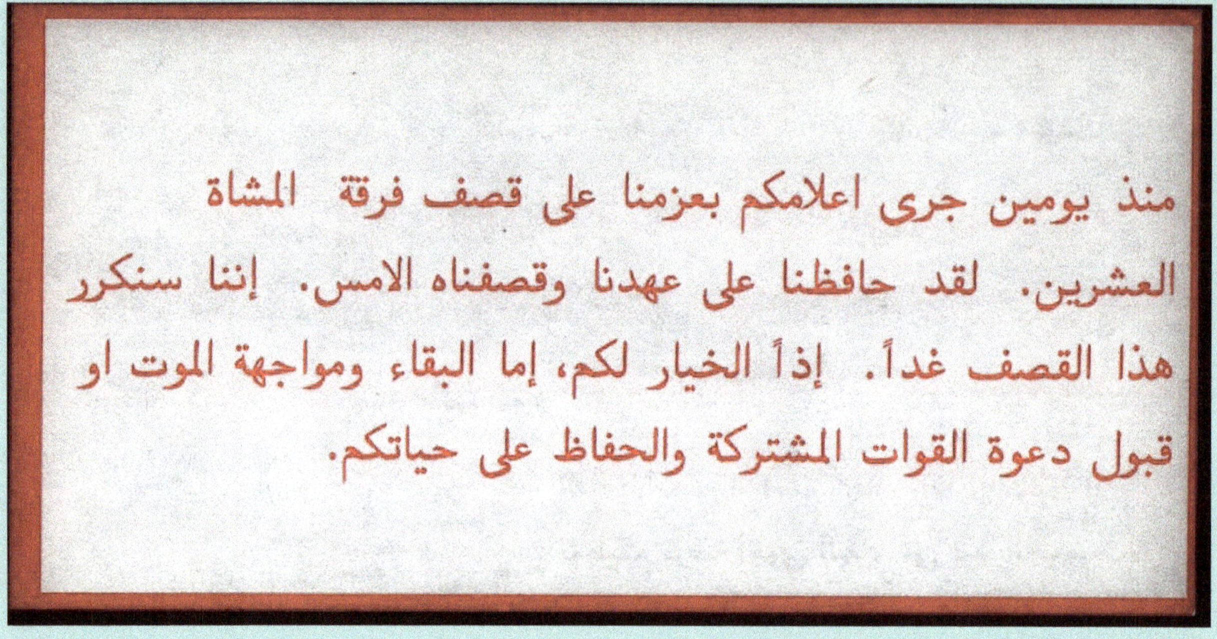

منذ يومين جرى اعلامكم بعزمنا على قصف فرقة المشاة
العشرين. لقد حافظنا على عهدنا وقصفناه الامس. إننا سنكرر
هذا القصف غداً. إذاً الخيار لكم، إما البقاء ومواجهة الموت او
قبول دعوة القوات المشتركة والحفاظ على حياتكم.

Two days ago, you were informed of our intention to bomb the 20th Infantry Division. We kept our promise and bombed it yesterday. We will repeat this bombing tomorrow. So, the choice is yours: either stay and face death or accept the call of the Joint Forces and preserve your lives.

Arabic

Leaflet 66

من أجل سلامتكم

* عند رؤيتكم سفن أو طائرات الإئتلاف وجهوا مقدمة سفينتكم
نحو العراق و سلموا سفينتكم او إجنحوا بها.

* إحشدوا الطاقم فوق ظهرها.   * إرفعوا علما او راية بيضاء فوق أعلى سارية.

* لا تأتوا بحركات تهديد.   * ضعوا كل الأسلحة بوضع اللارمي و غطوها.

* حاولوا الإتصال بسفن الإئتلاف بواسطة: القناة ١٦ من متن الى متن (١٥٦,٨ ميغاهرتز).
أو الطوارئ الجوية العالمية (١٢١,٥ ميغاهرتز).   أو الطوارئ الجوية الحربية (٢٤٣,٠٠ ميغا هيرتز).

باتباعكم إرشادات الإئتلاف يمكنكم البقاء احياء .

For Your Safety

When you see coalition ships or aircraft, steer your vessel towards Iraq and surrender your ship or beach it.

Assemble the crew on deck.

Raise a flag or a white banner atop the highest mast.

Do not make any threatening moves. Place all weapons in a non-firing position and cover them.

Try to contact coalition ships using Channel 16 for ship-to-ship communication (156.8 MHz), or the global air emergency frequency (121.5 MHz), or the military air emergency frequency (243.00 MHz).

By following coalition instructions, you can stay alive.

Arabic

Leaflet 67

بما أنكم خط الدفاع الأخير بين صدام و قوات الإئتلاف

فستواجهكم قوة ساحقة. لا تشاركوا صدام و نظامه نفس المصير.

لا تقاوموا قوات الإئتلاف.

لا تجعلوا مصيركم مرتبطا بمصير النظام.

Since you are the last line of defense between Saddam and the coalition forces, you will face overwhelming force. Do not share Saddam's fate and that of his regime.

Do not resist the coalition forces.

Do not tie your fate to the fate of the regime.

Arabic

Leaflet 68

Leaflet 69

Leaflet 70

No! Not for you!

Are these Arab customs and traditions?

Arabic

Leaflet 71

We can live for 20 years.

But our warehouses are empty, and your warehouses are full.

Arabic

Leaflet 72

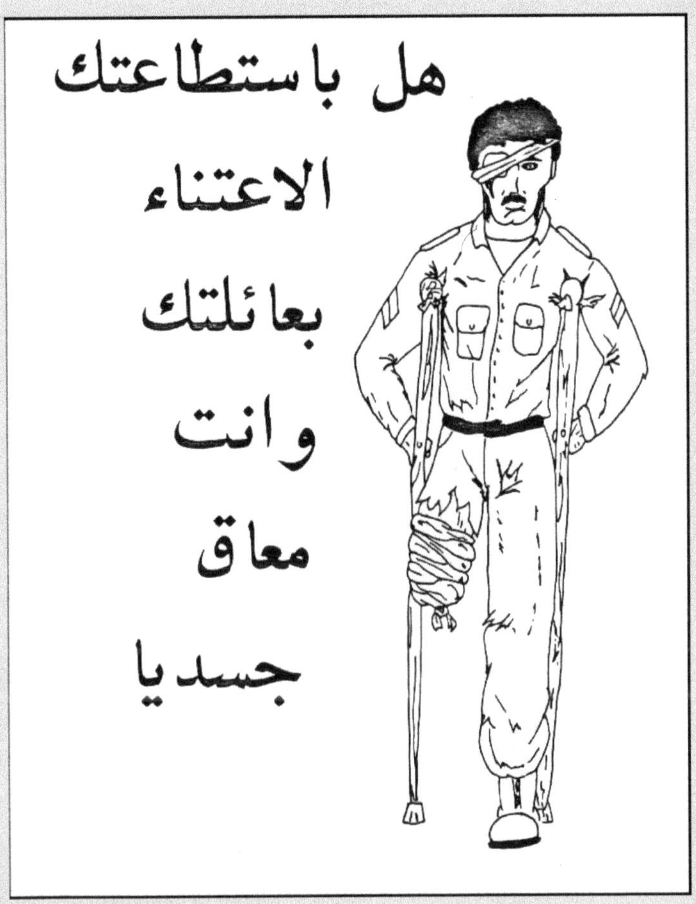

Can you take care of your family when you are physically disabled?

Arabic

Leaflet 73

Leaflet 74

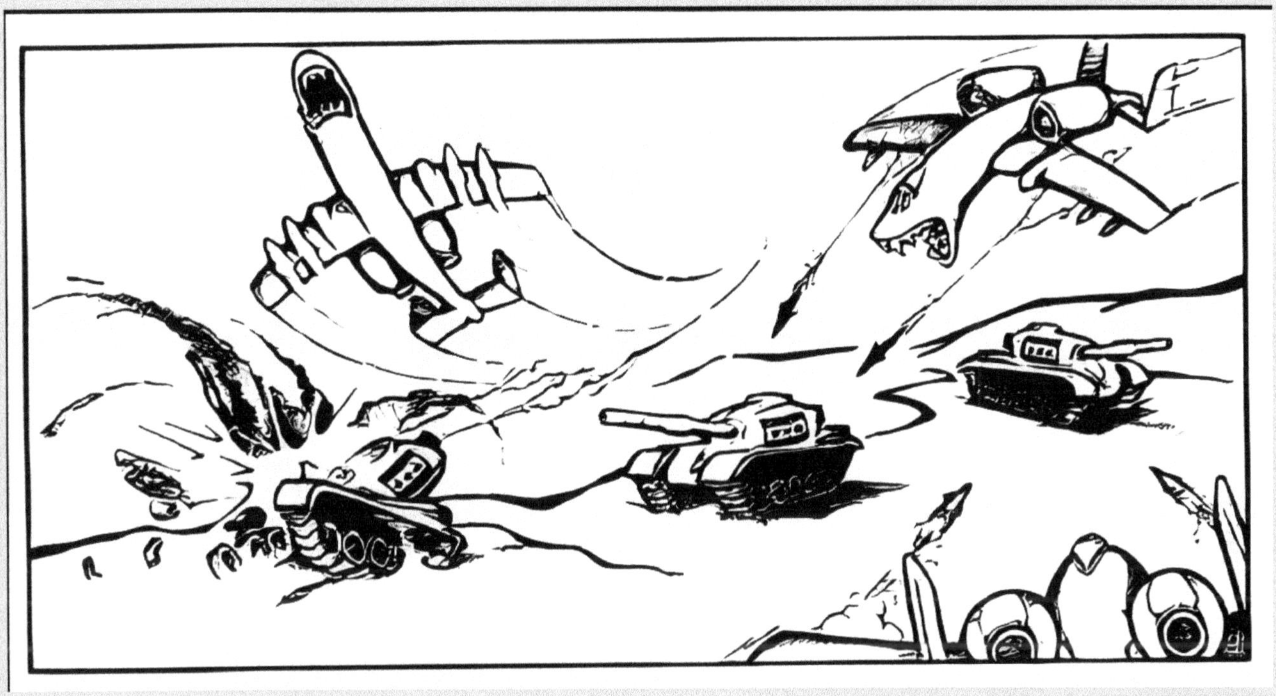

Leaflet 75

Leaflet 76

The foundation on which Saddam relies for power is on the verge of collapse.

Arabic

Leaflet 77

"At least it has value now."

Arabic

Leaflet 78

Green creep

The executioner of Baghdad

Arabic

Leaflet 79

The meal that Saddam gives you ?

List of daily meals for the coalition forces

Breakfast: Eggs, bread with butter, 2 fruits, fruit juice, milk, coffee, tea

Lunch: Meat, beans, or potato sandwich, cheese, sweets, fruit juice

Dinner: Meat, bread, vegetables, cake, milk, fruit and fruit juice, coffee, tea

Even the prisoners eat with the coalition soldiers during the mentioned meals.

Arabic

Leaflet 80

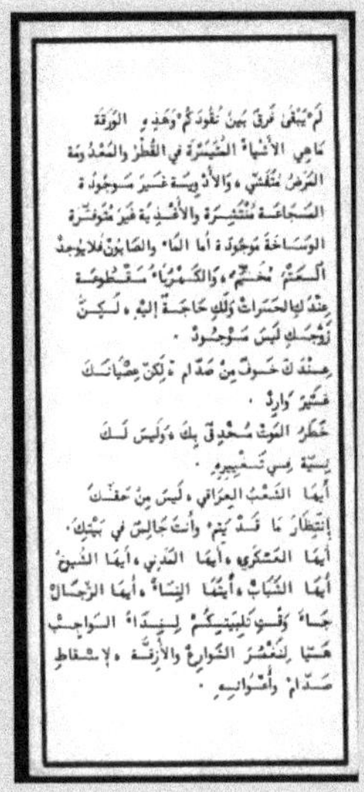

لم يبق فرق بين نقودكم وهذه الورقة
ما هي الأشياء المتيسرة في القطر والمعدومة
المرض متفشي ، والأدوية غير موجودة
المجاعة منتشرة والأغذية غير متوفرة
الوساخة موجودة أما الماء والصابون فلا يوجد
البكاء محتم ، والكهرباء مقطوعة
عندك الحسرات ولك حاجة إليه ، لكن
زوجك ليس موجود .
عندك خوف من صدام ، لكن عصيانك
غير وارد .
خطر الموت محدق بك ، وليس لك
نية في تغييره .
أيها الشعب العراقي ، ليس من حقك
إنتظار ما قد يتم وأنت جالس في بيتك .
أيها العسكري ، أيها المدني ، أيها الشيخ
أيها الشباب ، أيتها النساء ، أيها الرجال
جاء وقت تلبيتكم لنداء الواجب
هيا لننشر الشوارع والأزقة ، لإسقاط
صدام وأعوانه .

There is no longer any difference between your money and this piece of paper. What are the essentials available in the country? Food is scarce, medicine is unavailable, and famine is widespread.

Hygiene supplies are non-existent, and soap and water are nowhere to be found. You have the desire to cry, but you don't have the electricity to power a fan. You need it, but your husband isn't there.

You are afraid of Saddam, but your disobedience is a real risk. Death is looming over you, and you have no intention of changing anything.

Oh, people of Iraq, you have no right to wait for what may come while you sit in your house. Soldiers, civilians, the elderly, the youth, women, men, the time has come for you to heed the call of duty.

Fill the streets and alleys, and overthrow Saddam and his followers.

Arabic

Leaflet 81

أنّ النقود في العراق، أصبحت ليست لها قيمة .

حيث نقص في المواد الغذائية ، وليس هناك راحة ولا السّعادة .

أن في العراق اليوم ، عطش و مجاعة و سفالة و موت .

أن صدّام حسين، هو المسؤول عن ذلك كله .

والآن باشر بالعمل ضدّ نظام صدّام . . . لابد من سقوط صدّام .

The money in Iraq has lost its value.
There is a shortage of food supplies, and there is no comfort or happiness.
In Iraq today, there is thirst, famine, misery, and death.
Saddam Hussein is responsible for all of this.
Now, start acting against Saddam's regime... Saddam must fall.

Arabic

Leaflet 82

The 16th Infantry Division will be bombed tomorrow. Leave this position now and save yourself.

Arabic

Leaflet 83

سبق ان اخبرناكم بعزمنا على قصف الفرقة السادسة عشر (١٦)
المشاة ولقد حافظنا على هذا الوعد وتم قصفها بالامس

احذروا

سنكرر القصف مرة اخرى غداً .... الان اصبح الخيار لكم
اما البقاء ومواجهة الموت او قبول دعوة القوات المشتركة للحفاظ على حياتكم

We have already informed you of our promise to bomb the 16th Infantry Division. We kept our promise and bombed them yesterday.

BEWARE.

We will repeat this bombing tomorrow. Now the choice is yours. Either stay and face death or accept the invitation of the Joint Forces to protect your lives.

Arabic

Leaflet 84

I have carried you on my back for eleven years...
how can I have any strength left now?

Arabic

Leaflet 85

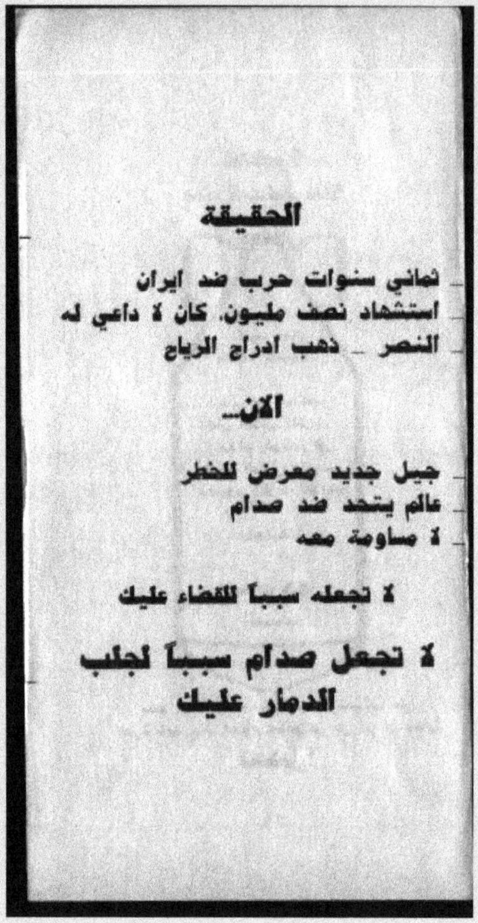

الحقيقة

ـ ثماني سنوات حرب ضد ايران
ـ استشهاد نصف مليون. كان لا داعي له
ـ النصر _ ذهب ادراج الرياح

الآن..

ـ جيل جديد معرض للخطر
ـ عالم يتحد ضد صدام
ـ لا مساومة معه

لا تجعله سبباً للقضاء عليك

لا تجعل صدام سبباً لجلب الدمار عليك..

The truth
Eight years of war against Iran, half a million were martyred, the victory was unnecessary - it was in vain now...
A new generation at risk, a world uniting against an uncompromising Saddam
Don't let it be a reason to eliminate you.
Do not let Saddam bring destruction to you.

Arabic

Leaflet 86

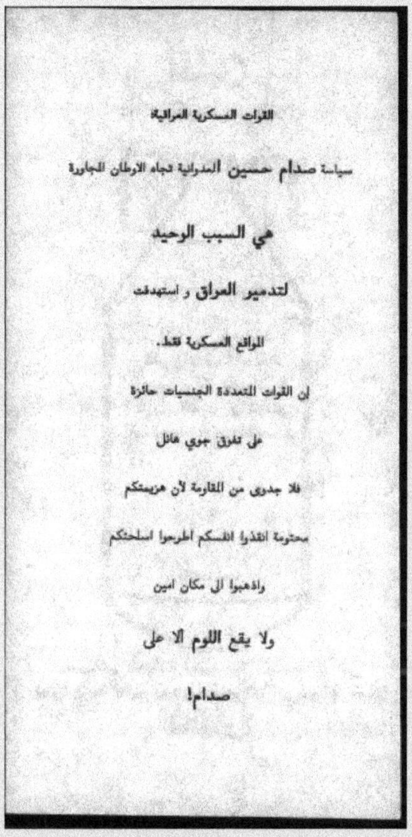

القوات العسكرية العراقية

سياسة صدام حسين العدوانية تجاه الاوطان المجاورة

هي السبب الوحيد

لتدمير العراق و استهدفت

المواقع العسكرية فقط.

إن القوات المتعددة الجنسيات حائزة

على تفوق جوي هائل

فلا جدوى من المقاومة لأن هزيمتكم

محتومة انقذوا انفسكم اطرحوا اسلحتكم

واذهبوا الى مكان امين

ولا يقع اللوم الا على

صدام!

Iraqi military forces:

Saddam Hussein's policy of aggression is the only reason for the bombing of Iraq. The bombing is for military targets only. The multi-national air forces have overwhelming air superiority. Resistance is useless. The outcome is inevitable. Save yourselves. Leave your weapons and immediately go to a safe area.

Saddam is to blame!

Arabic

Leaflet 87

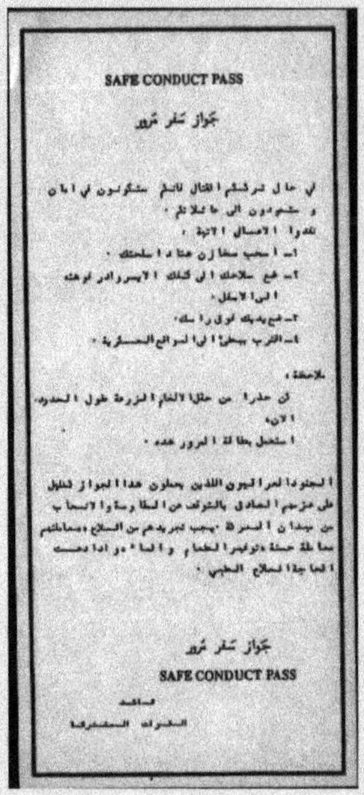
SAFE CONDUCT PASS

SAFE CONDUCT PASS

SAFE CONDUCT PASS.

You do not have to die.

You can be safe and return to your family and loved ones if you cease resistance. You must follow these steps strictly:

1: Remove the magazine from your weapon.

2: Sling your weapon over your left shoulder with the muzzle down.

3: Raise your arms above your head.

4: Approach positions slowly. Hold this pass in your hand above your head. If you do this, you will not die. You will be treated well and someday return to your family.

NOTE: Beware of mine fields along the border.

The Iraqi soldiers that carry this safe conduct pass are using it as a sign of their sincere desire to cease resistance and be removed from the battlefield.

They are to be disarmed, to be well treated, to receive food, water, and medical treatment if necessary.

Safe Conduct Pass.

Commander,

Combined Forces.

Arabic

Leaflet 88

Safe conduct card

The bearer of this card is permitted to cross the border to the coalition forces, whether allied or friendly, especially the American, British, and French forces. He will receive good treatment from the joint forces near the headquarters of the coalition forces. He will be treated safely and securely and without harm, in accordance with the Geneva Convention.

Commander
Joint Forces and Theater of Operations

Arabic

THE WAR IN AFGHANISTAN

War on Terrorism

(October 7, 2001 – August 30, 2021)

Leaflet 89

There is no chance of escape.

Pashto, Arabic

Leaflet 90

Al-Qaeda, your escape routes are mined.

Pashto, Dari

Leaflet 91

A killer being wanted

No place to hide, no killer escapes

Wanted for committing murders: Osama bin Laden

Arabic

Leaflet 92

Escape to survive.

Pashto, Dari

Leaflet 93

Do not touch the mines.

The Danger!

Mine!

Pashto, Dari

Leaflet 94

The Taliban/ Al-Qaeda members you are the target of our attacks.

Pashto, Dari

Leaflet 95

But why do you give safe passage to people
who betrayed you and left you alone.

Pashto

Leaflet 96

It is terrifying. These are people who sent your brothers and sisters to the war to die.

Pashto

Leaflet 97

Osama bin Laden, the murderer and coward, has left you. He abandoned you and fled. Surrender before you get killed. Bin Laden is not worth dying for. Do not let your death bring sorrow and grief to your family.

Pashto, Dari

Leaflet 98

Hamid Karzai
The head of the interim government of Afghanistan.

Pashto, Dari

Leaflet 99

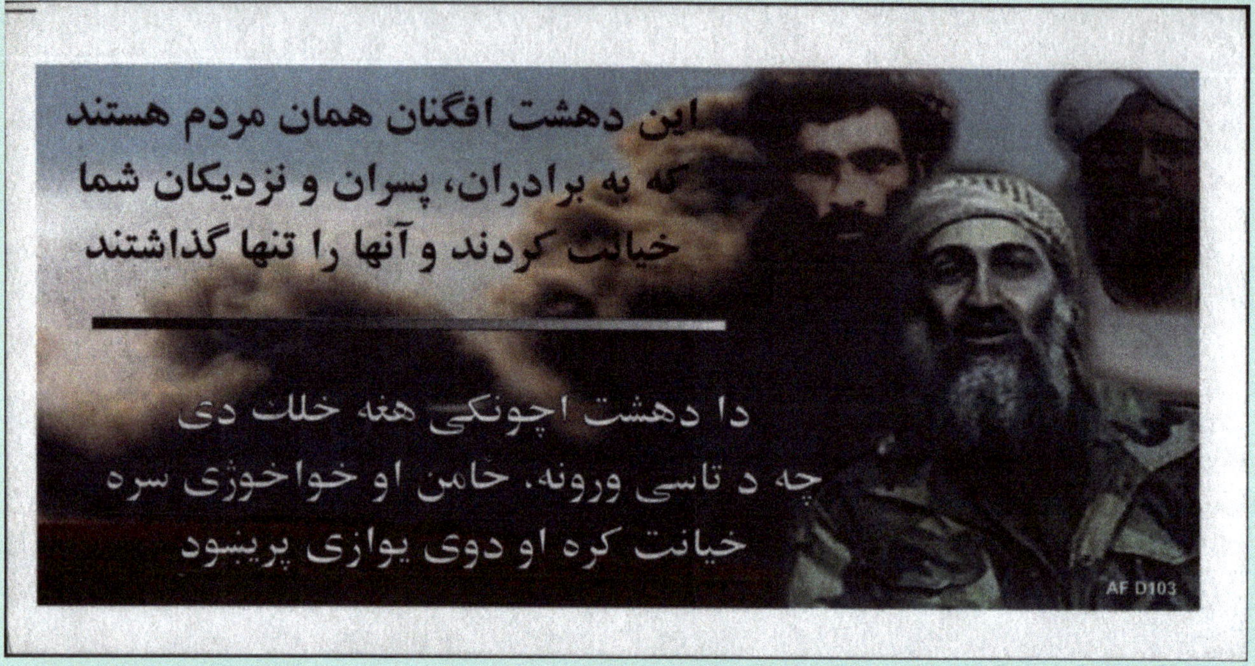

Terrorists are the people who do not care about
your family or your life, t hey are traitors.

Pashto, Dari

Leaflet 100

Why do you let these people take your brothers away to fight when they do not know why they are fighting or what they are fighting for?

Pashto, Dari

Leaflet 101

Taliban and Al-Qaeda brought pain and trouble to Afghanistan.

Pashto, Dari

Leaflet 102

Taliban and Al-Qaeda members, surrender to the coalition forces.

Pashto, Dari

Leaflet 103

Life with terrorists is a life of fear.

Pashto

Leaflet 104

A life without terrorists is a life of peace and prosperity.

Pashto

Leaflet 105

Osama bin Laden the honorless murderer has abandoned you.

Pashto, Dari

Leaflet 106

Zawahiri is an honorless murderous man.

Pashto

Leaflet 107

There is no possibility of escape and hiding.

Pashto, Dari

Leaflet 108

Al Qaeda and the Taliban have destroyed your cities.

Pashto, Dari

Leaflet 109

Stop the fight!

Pashto

Leaflet 110

Afghanistan is an Islamic Republic, not a Mullahcracy.

Pashto

Leaflet 111

Reward

Osama bin Laden

Ayman Al-Zawahiri

25 million dollars

Pashto, Dari

Leaflet 112

Up to 25 million dollars reward for information about the location of these men.

Pashto, Dari

Leaflet 113

The attack of the coalition forces prevents all humanitarian aid from reaching your region.

Pashto, Dari

Leaflet 114

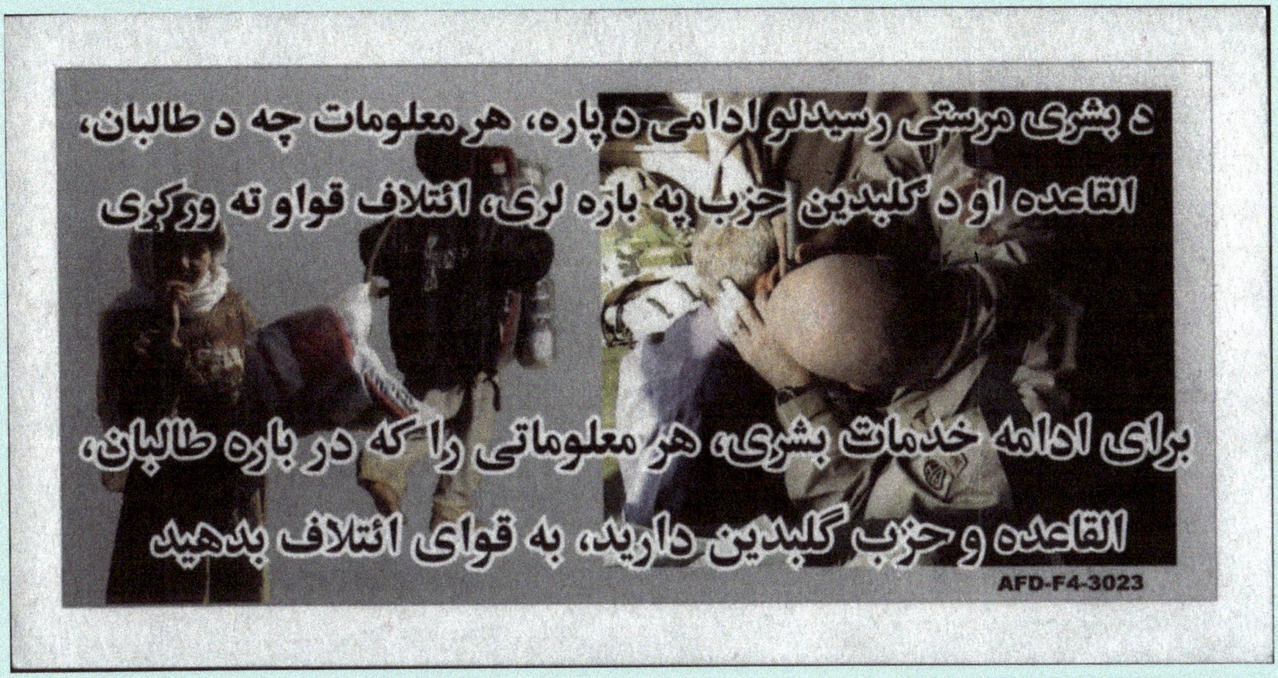

To continue recieving humanitarian services, give any information you have about Taliban, Al-Qaeda and Gulbuddin Hizb to the coalition forces.

Pashto, Dari

Leaflet 115

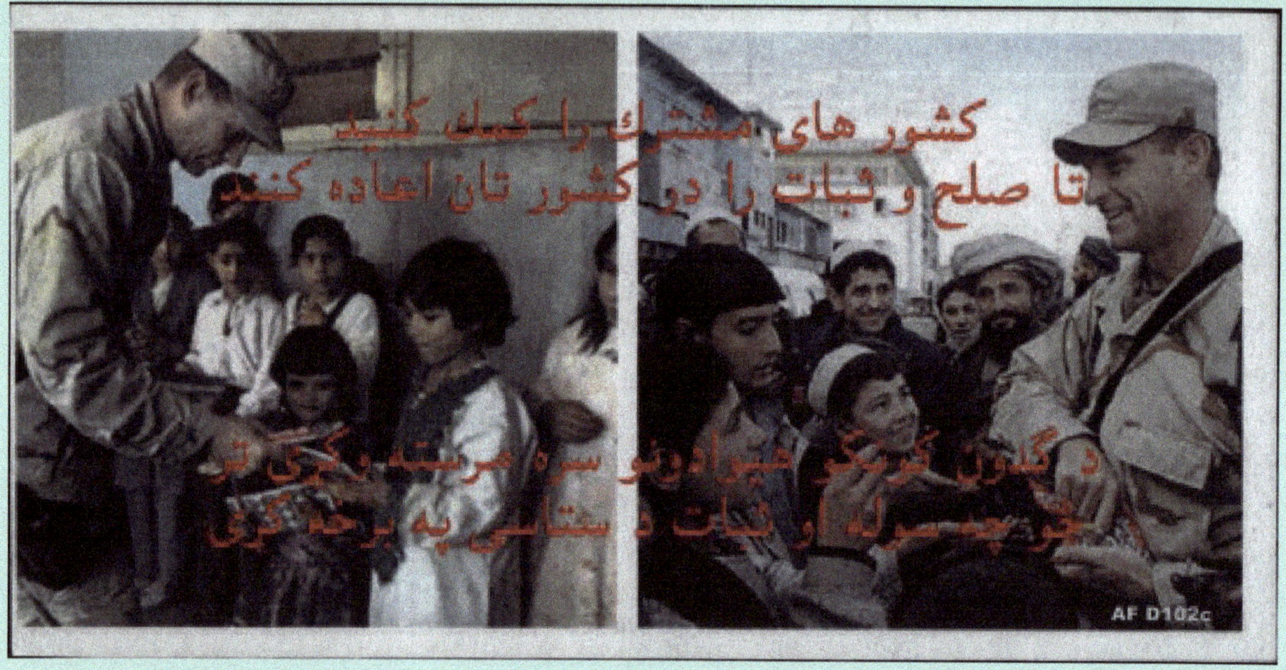

Help the Commonwealth Nations to restore peace and stability in your country.

Pashto, Dari

Leaflet 116

For your protection, do not interfere with the personnel and equipment of the host countries.

Pashto, Dari

Leaflet 117

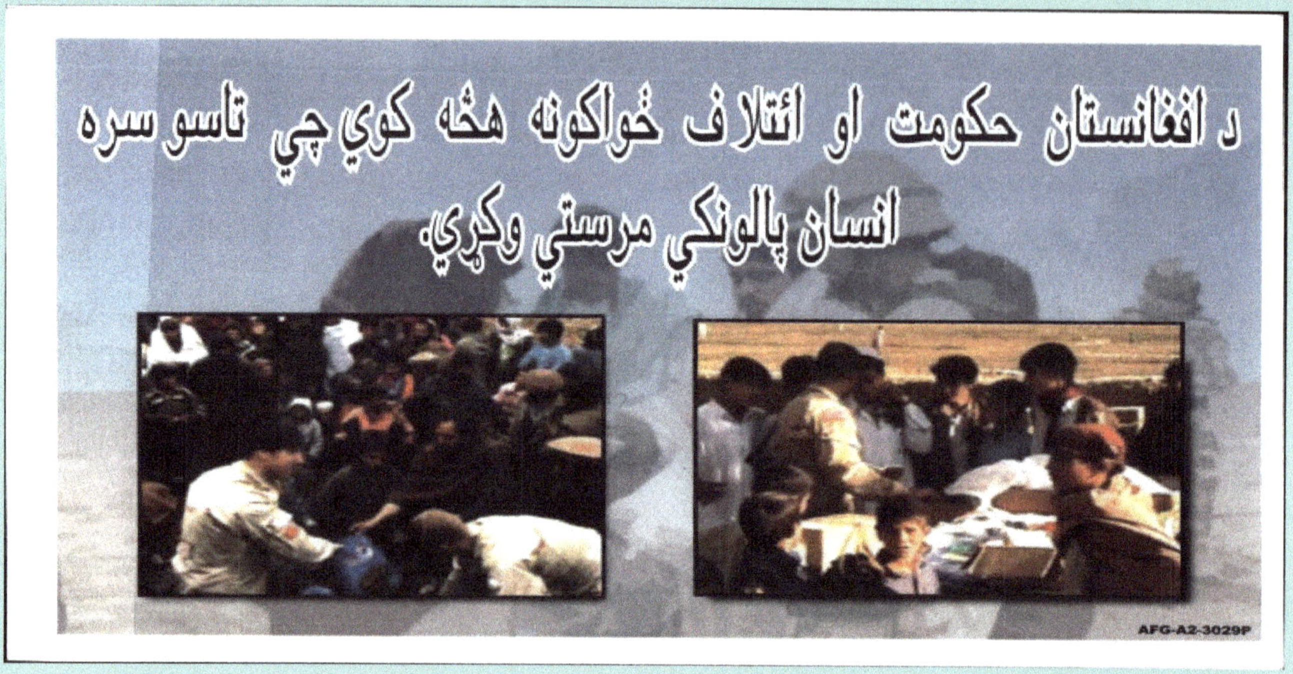

The government of Afghanistan and the coalition forces are trying to help you With humanitarian assistance.

Pashto

Leaflet 118

Attacks on humanitarian organizations and coalition countries deprive you and your family of the help you need. Stop attackers from attacking coalition forces by reporting them.

Pashto, Dari

Leaflet 119

The smuggling of Iraq's natural materials is the theft of people's wealth and will be punished.
Traffickers will be brought to justice in Iraq and severely punished.
Central Criminal Court convicts oil smuggler!
No one can escape from the law. It is the duty of the people to inform the coalition forces about illegal activities.

Pashto, Dari

Leaflet 120

Name: Mozerenko Mikola

Crime: Oil smuggling

Order of Justice: Criminal

Penalty: Seven years in prison

Fine: $2.40 million

He is trying to find himself in prison.

Name: Ivan Soschenko

Crime: Oil smuggling

Order of Justice: Guilty

Penalty: Seven years imprisonment

Fine: $2.4 million

Pashto, Dari

Leaflet 121

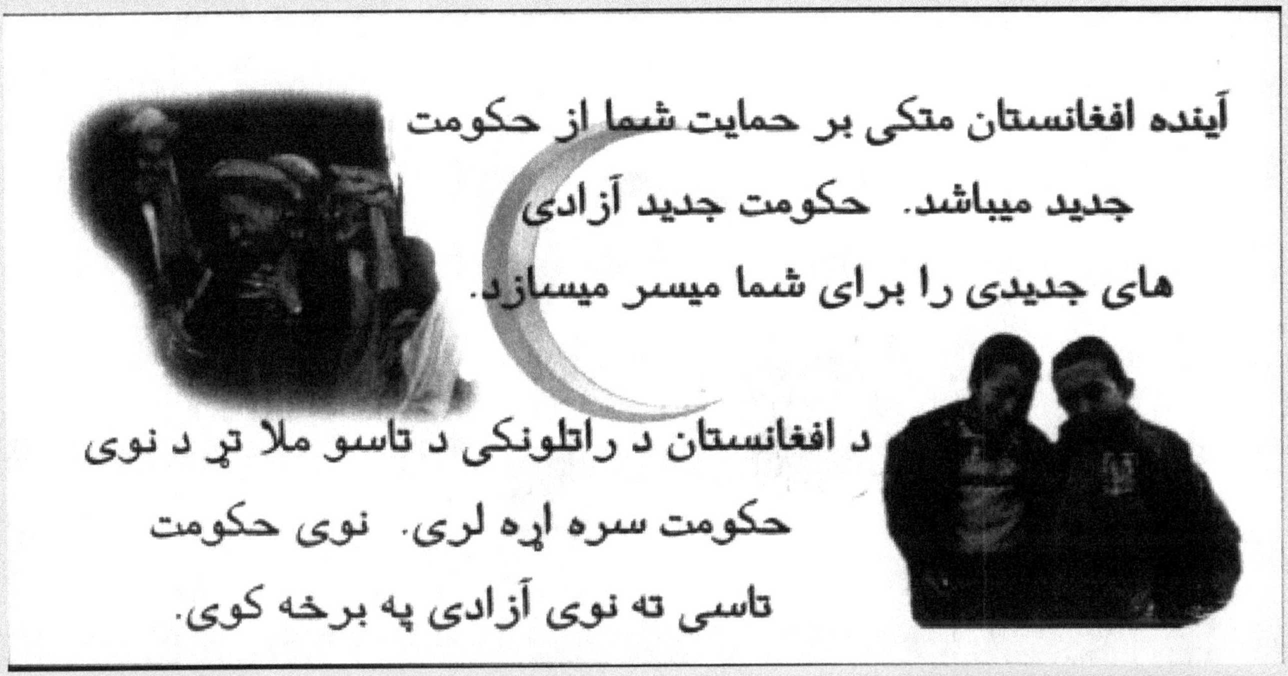

The future of Afghanistan depends on your support for the new government.
The new government gives you new freedoms.

Pashto, Dari

Leaflet 122

Stay and die.

Pashto, Dari

Leaflet 123

By attacking Afghanistan, death is inevitable for you.

Pashto, Dari

Leaflet 124

We stop your vehicles because the enemies of Afghanistan have been seen in this area.

The enemies of Afghanistan have attacked the defenseless people of Afghanistan and Coalition forces. We are here to help the peace-loving people and to chase, Capture or kill the criminals.

If you have any information about the enemy, please inform the local Governor, local police, the Afghan National Army or Coalition forces and take part in bringing permanent peace to your country.

Pashto, Dari

Leaflet 125

America has helped Afghanistan with 170 million dollars.

Pashto, Dari

Leaflet 126

The enemies of Afghanistan cannot stop your children from laughing and playing every day. The children of Afghanistan will grow up and make Afghanistan proud. For the sake of a bright future for Afghanistan, inform the coalition forces about the activities of the militias.

Pashto, Dari

Leaflet 127

The enemies of Afghanistan can no longer control the villages.

Pashto, Dari

Leaflet 128

You cannot hide!

Pashto

Leaflet 129

The Coalition forces can provide a stable environment to allow Afghanistan to rebuild.

Pashto, Dari

Leaflet 130

It is your national duty to inform the National Army and Coalition Forces about the terrorists.

Pashto, Dari

Leaflet 131

The future of Afghanistan is in your hands.

Pashto, Dari

Leaflet 132

Return Afghanistan to its former glory. Help to restore Afghanistan's past.

Pashto, Dari

Leaflet 133

Afghan Nationals, the army has come to provide you security.

As an Afghan national, you are responsible for the security of your family and Afghanistan.

Pashto, Dari

Leaflet 134

The Afghan National Army is here to help rebuild Afghanistan.

Pashto, Dari

Leaflet 135

The soldiers of the Afghan National Army are working for a better future for all Afghans.

Pashto, Dari

Operation Enduring Freedom

(October 7, 2001 – December 28, 2014)

Leaflet 136

Leaflet 137

Leaflet 138

Al-Qaeda, do you worry if you are safe?

Pashto, Dari

Leaflet 139

In your graves?

Pashto, Dari

Leaflet 140

Award up to 25 million dollars.

Osama bin Laden

Pashto, Dari

Leaflet 141

Up to 25 million dollars reward for providing reliable information about the whereabouts or capture of Osama bin Laden.

Pashto, Dari

Leaflet 142

Osama bin Laden
Ayman Al-Zawahiri
Award up to 2.5 million dollars.

Pashto, Dari

Leaflet 143

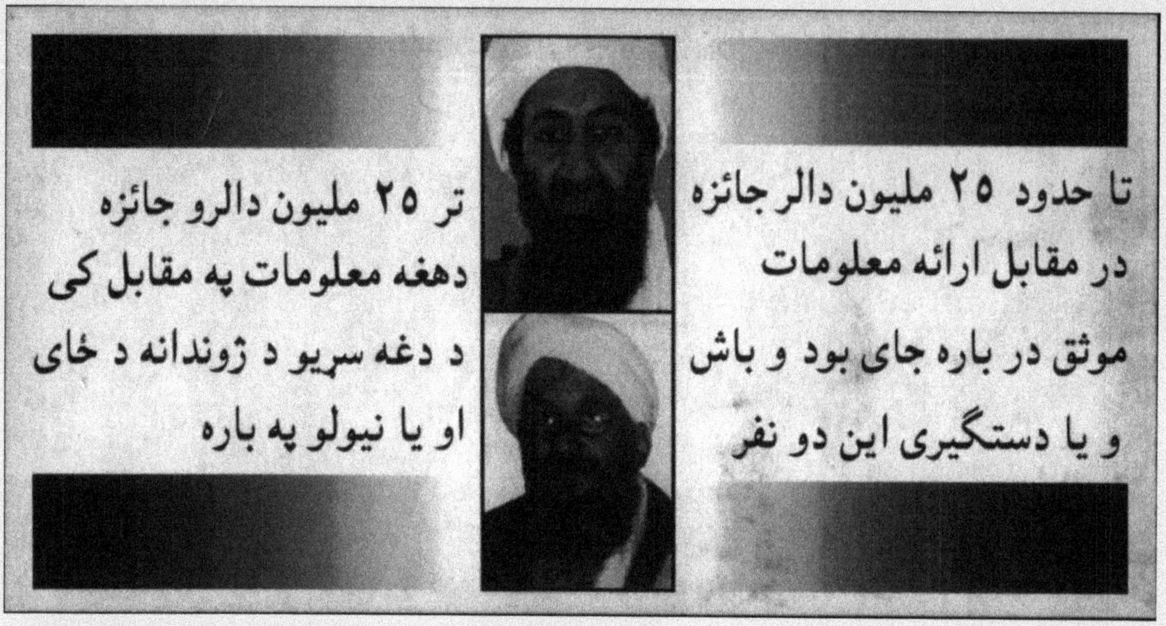

Up to $25 million in reward for information about the whereabouts or capture of these men.

Pashto, Dari

Leaflet 144

The terrible government of the Taliban.

Haqqani

Bin Laden

Mutawakil

Pashto, Dari

Leaflet 145

It is falling now!

Bin Laden

Haqqani

Mutawakil

Pashto, Dari

Leaflet 146

Osama bin Laden laughs at you as you do not know he has sent you to your death.

Pashto, Dari

Leaflet 147

Osama bin Laden sends his murderers into the world to kill for his cause.

Pashto, Dari

Leaflet 148

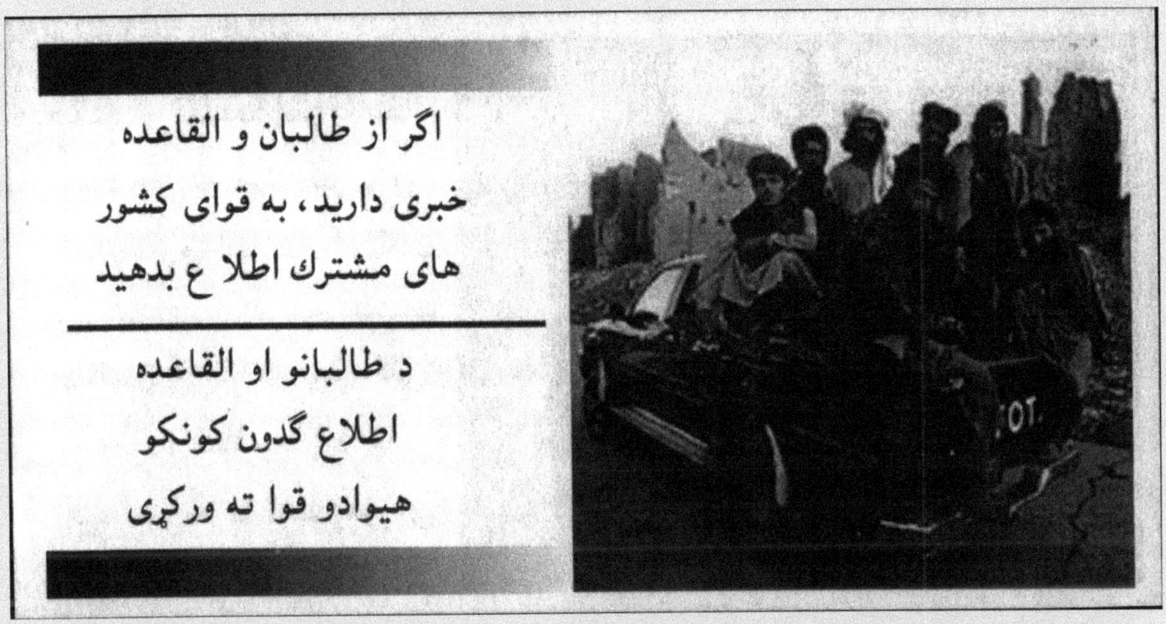

If you have information about the Taliban or Al-Qaeda, inform the coalition forces immediately.

Pashto, Dari

Leaflet 149

Taliban and Al-Qaeda use innocent women and children as shields for protection.

Pashto, Dari

Leaflet 150

Your children are waiting for you. For peace, stability and a bright future.

Pashto, Dari

Leaflet 151

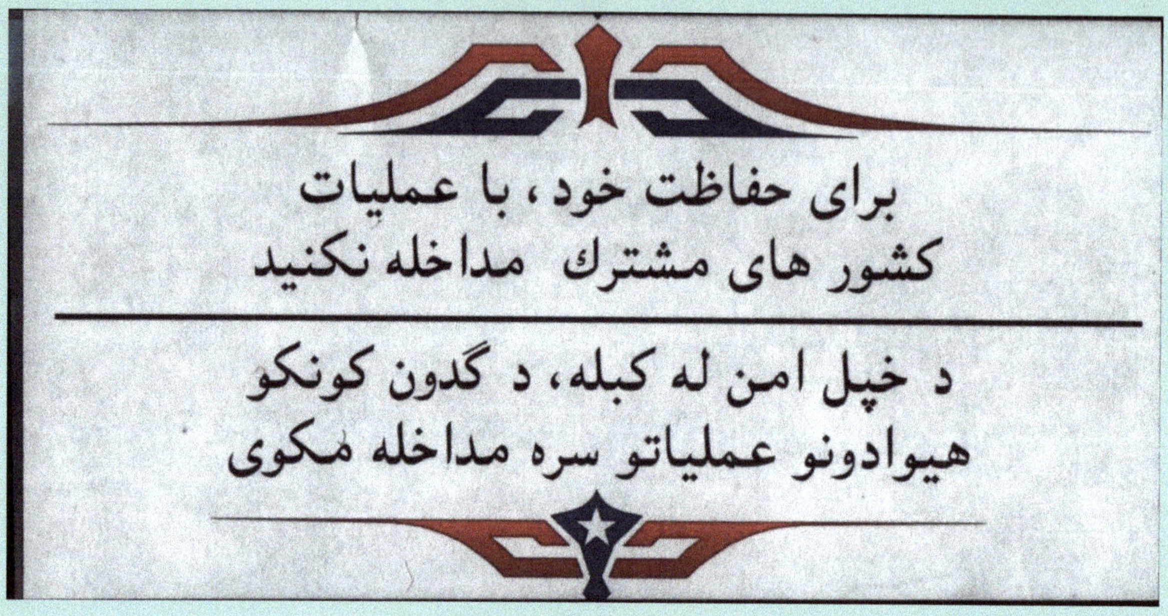

To protect yourself, do not interfere with the operations of NATO countries.

Pashto, Dari

Leaflet 152

که تاسی د طالبان ضد قوا سره مرسته وکړی چه د القاعده او طالبی قاتلا ن
ونیول شی، د میلیونو ډالرو چشتن کیدلا ی شی. دا پانګه دومره زیات ده چه
د تاسی د فامیل، کلی او قوم د پاره تر آخره ژوند پوری بس دی. د دغه پیسی
سره تاسی کولا ی شی چه د مکتب کتابونه، غوا، غوای او پسونه واخلی، د
ډاکتر اجورت ورکړی او هم خپل ټول کلی ته کورونه جوړ کړی.

If you help the forces against the Taliban and Al-Qaeda and capture them, you will receive a reward of millions of dollars. This amount is more than enough for you and your family for the rest of your lives. With this money, you can buy books, animals, and land, and build a house for yourself and your children. You will also be provided with a job and other benefits.

Pashto

Leaflet 153

Report the presence of explosive materials to prevent attacks against minors. The terrorists' campaigns use improvised explosive devices. Inform the coalition forces about the locations of new minefields.

Pashto

Leaflet 154

Mass murder all over the world

Hundreds of civilians were killed in Nairobi.

20 innocent people from 9 countries including 2 children and 10 Arabs were killed in Riyadh.

More than 200 innocent people from 15 countries have been killed or disappeared in Bali.

23 international humanitarian workers were killed in the United Nations building in Baghdad.

Help so that innocent people are not killed in the future.

Contact the embassies of the allied countries or the members of the navy.

Pashto, Dari

Leaflet 155

Coalition forces are looking for Al-Qaeda members and other terrorists. If you are aware of their plans or terrorist activities, report them to these locations.

Do it:

To a member of the Navy

To coalition embassies

US Awards program for Justice (0118008773927)

By phone number or email (mail@dssrewards.net).

Your identity will be protected, and your family will be protected.

Pashto

Leaflet 156

Observe healthy conditions in daily life.
Choose a healthy way of life.
This water has germs and microbes that can make you sick or even die.
It contains germs and bacteria that cause your illness and even death.
To eliminate germs and bacteria in water, it should be boiled for 10 minutes before

Pashto, Dari

Leaflet 157

You cannot hide from the water.

If you observe someone placing a mine or know about someone placing a mine in the water, inform the coalition forces immediately. You will receive a substantial reward. Your identity will be kept confidential.

Your information can save lives.

Urdu

Leaflet 158

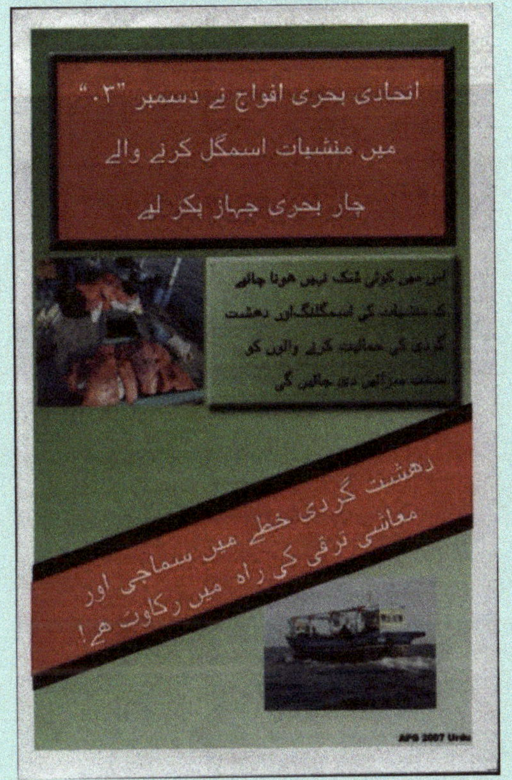

The coalition naval forces seized a ship smuggling drugs on December 3rd.

Do not support those who smuggle drugs for terrorism.

Instead, inform the coalition forces immediately.

Terrorism is a threat to peace and prosperity in the region!

Urdu

Leaflet 159

Information and news radio

0500-1000.

1700-2200 daily.

864, 1107, 8700 kilohertz.

Pashto, Dari

Leaflet 160

Syed Hussain Anwari

Sima Samar

Abdullah

Hamid Karzai

Support the interim government of Afghanistan.

Pashto, Dari

Leaflet 161

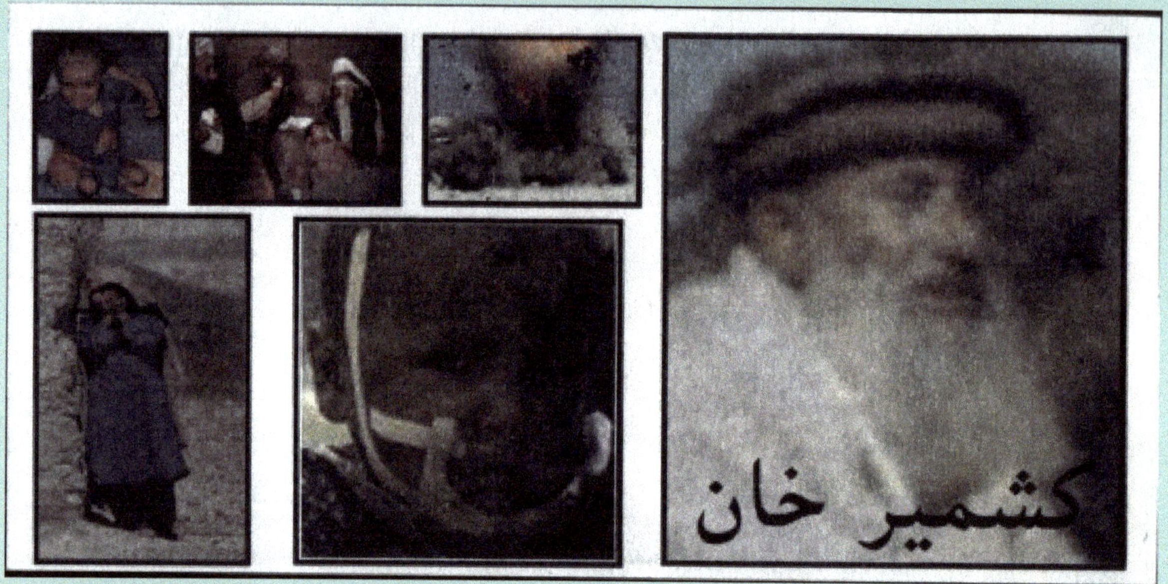

Kashmir Khan

Pashto, Dari

Leaflet 162

Support your interim government to ensure a better life.

Pashto, Dari

Leaflet 163

Taliban: We know your location!

Arabic

Leaflet 164

Stop fighting for the Taliban and you will stay alive!

Arabic

Leaflet 165

America has provided over $170 million in aid to Afghanistan.

Pashto, Dari

Leaflet 166

This is what the Taliban has done.

Pashto, Dari

Leaflet 167

Help the anti-Taliban forces to drive the murderers and thugs out of Afghanistan with the amount of money you never dreamed of.

Pashto, Dari

Leaflet 168

Many tribes make a united Afghanistan.

Pashto, Dari

Leaflet 169

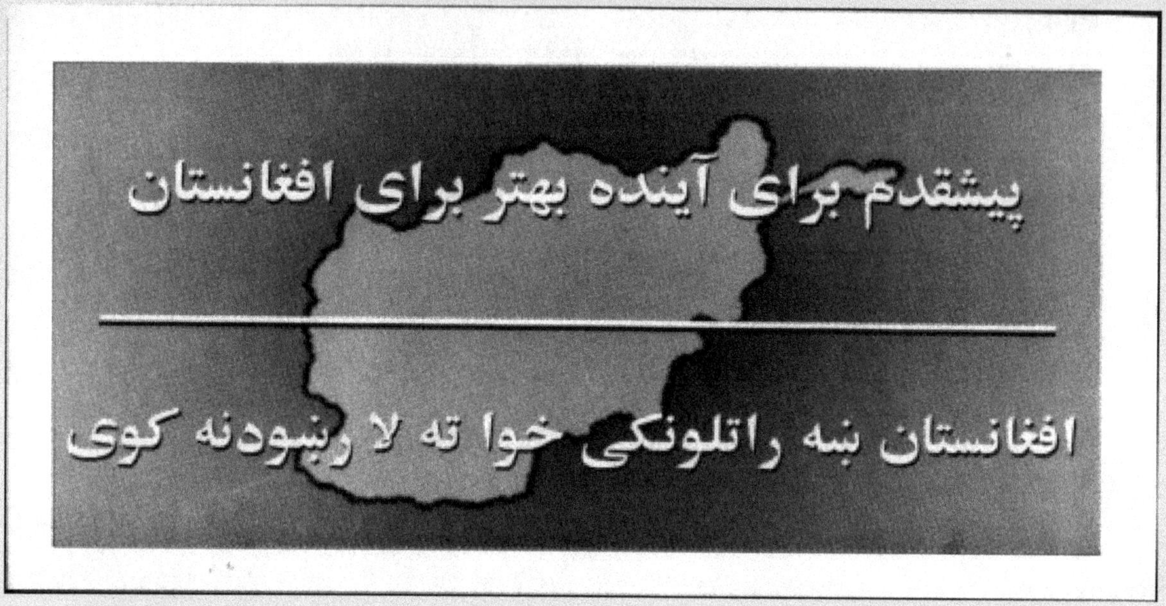

Afghanistan is heading towards a better future.

Pashto, Dari

Leaflet 170

Friendship

Afghanistan

The United States

Pashto, Dari

Leaflet 171

The United Nations has come here to help the people of Afghanistan.

Pashto, Dari

Leaflet 172

A united Afghanistan brings peace and prosperity to you.

Pashto, Dari

Leaflet 173

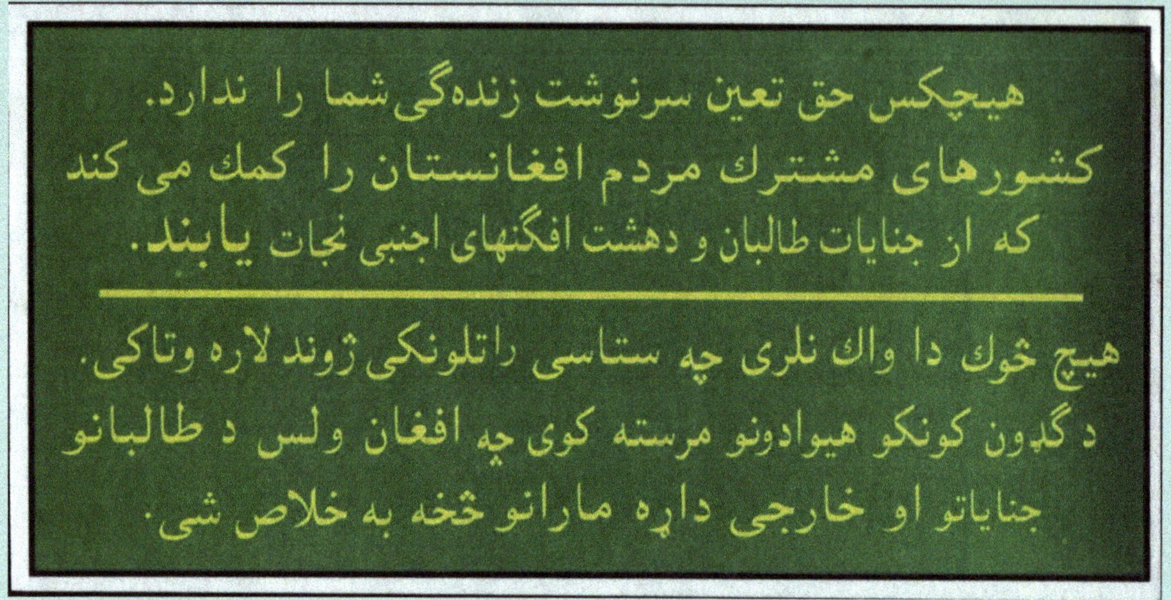

No one has the right to determine the fate of your life. NATO countries help the people of Afghanistan to save themselves from the crimes of the Taliban and foreign terrorists.

Pashto, Dari

Leaflet 174

A carpet is made from several yarns!

Pashto, Dari

Leaflet 175

Osama bin Laden and his foreign followers do not want the people of Afghanistan to live in peace with each other. Afghans must rid themselves of these fanatics.

Dari, Pashto

Leaflet 176

The time has come for Afghans to make peace among themselves.

Dari, Pashto

Leaflet 177

Drop by drop becomes the sea and brick by brick becomes the house.

Pashto, Dari

Leaflet 178

A united Afghanistan brings peace and prosperity.

Afghanistan

Pashto, Dari

Leaflet 179

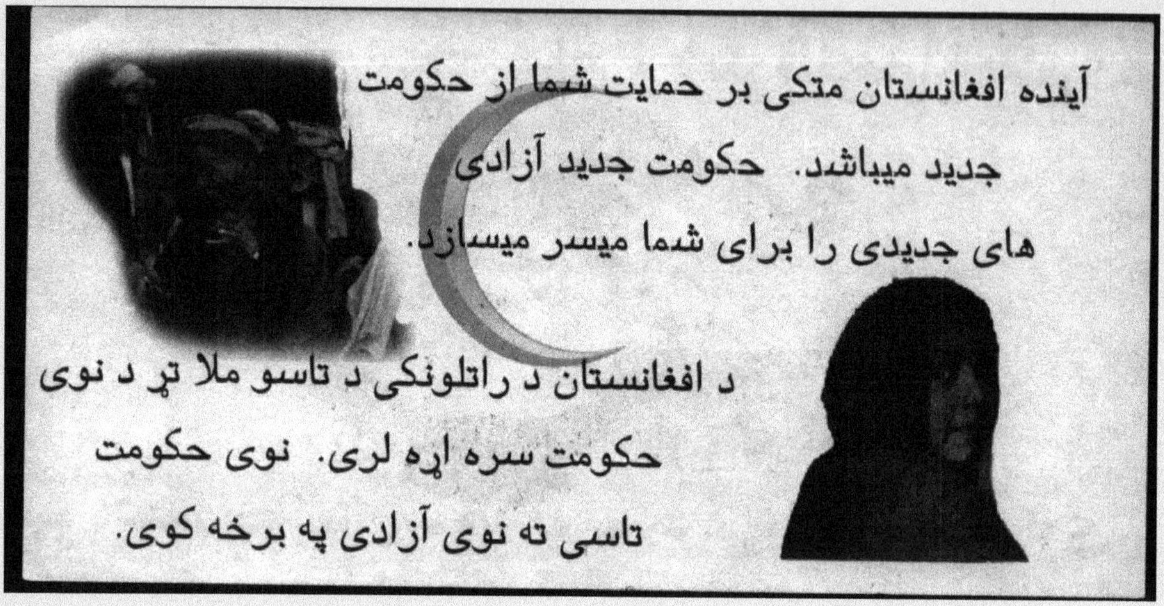

The future of Afghanistan depends on your support for the new government.
The new government gives you new freedoms.

Pashto, Dari

Leaflet 180

Drive the foreign owners out of the country and live in peace!

Bin Laden

Foreign masters and evil will live in peace!

Pashto, Dari

Leaflet 181

Who are the real administrators of the Talib regime?
Mullah Omar (on leash) and Bin Laden (holding leash)

Pashto, Dari

Leaflet 182

The Commonwealth Nations is here to help the people of Afghanistan.

Pashto, Dari

Leaflet 183

The Commonwealth is here to help you.

Pashto, Dari

FIGHT AGAINST ISIS-KHORASAN

(January 2015 – Ongoing)

Leaflet 184

Daesh, you are hunted by the Afghan government.

You have the chance to gain a prosperous life when you call the number 0795661142.

Pashto

THE IRAQ WAR

Operation Iraqi Freedom

(March 20, 2003 – December 15, 2011)

Leaflet 185

Iraq

The Coalition is here to put an end to the oppression caused by Saddam and his regime.

Arabic

Leaflet 186

Iraq

Military audio fiber cords are a tool used by Saddam and his regime in order to suppress the Iraqi people.

Arabic

Leaflet 187

Think about your family.
Act in a way that keeps you alive.

Arabic

Leaflet 188

Before coalition aircraft attack,
Think about the consequences.

Arabic

Leaflet 189

Do not use weapons of mass destruction.

Arabic

Leaflet 190

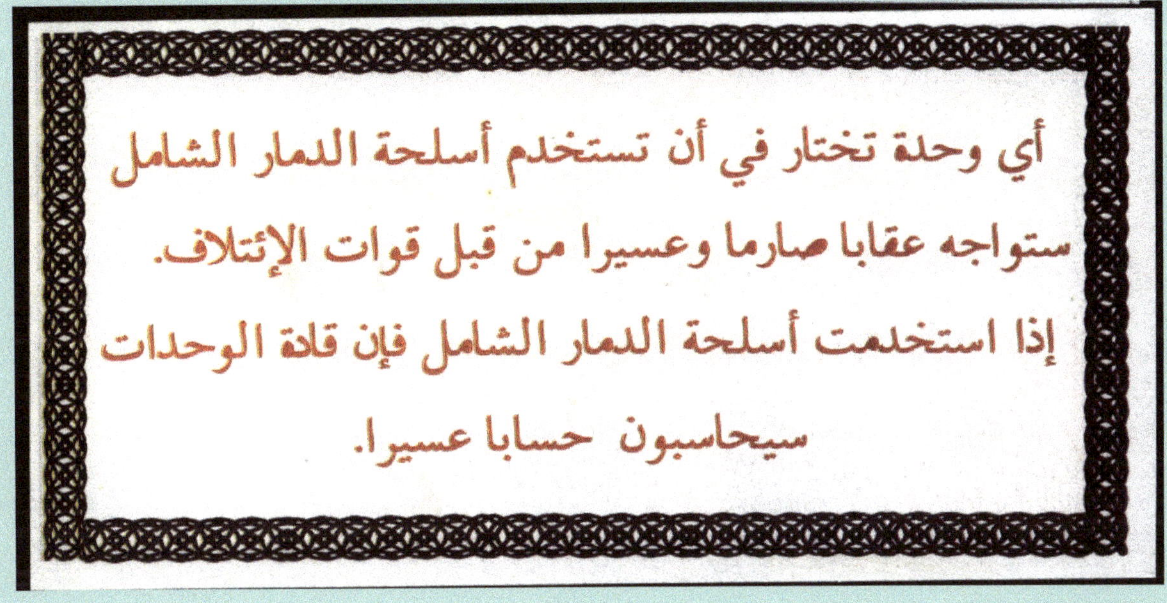

Any unit that chooses to use weapons of mass destruction will face severe and harsh consequences from the coalition forces. If weapons of mass destruction are used, the unit commanders will be held strictly accountable.

Arabic

Leaflet 191

يا شعب العراق، نرجو منكم البقاء في منازلكم آمنين. لا تذهبوا إلى أعمالكم ولا تقودوا
ليلا. ربما، أخطأت طائرات الإئتلاف وحسبت أن سياراتكم هي آليات عسكرية. من فضلكم لا
تحاولوا التدخل بعمليات الإئتلا ف حتى لا تصبحوا هدفا. لا تسمحوا لأولادكم باللعب في
الهواء الطلق. سوف تقدم إليكم تعليمات معينة فور بدأ قوات الإئتلاف بتوزيع المساعدات
الإنسانية.

People of Iraq, please stay home and stay safe. Do not go to work and do not drive at night. Perhaps, the coalition aircraft mistook your vehicles for military vehicles. Please do not try to interfere with the coalition's operations so that you do not become a target. Do not allow your children to play outdoors. Specific instructions will be provided to you once Coalition Forces begin distributing aid.

Humanity.

Arabic

Leaflet 192

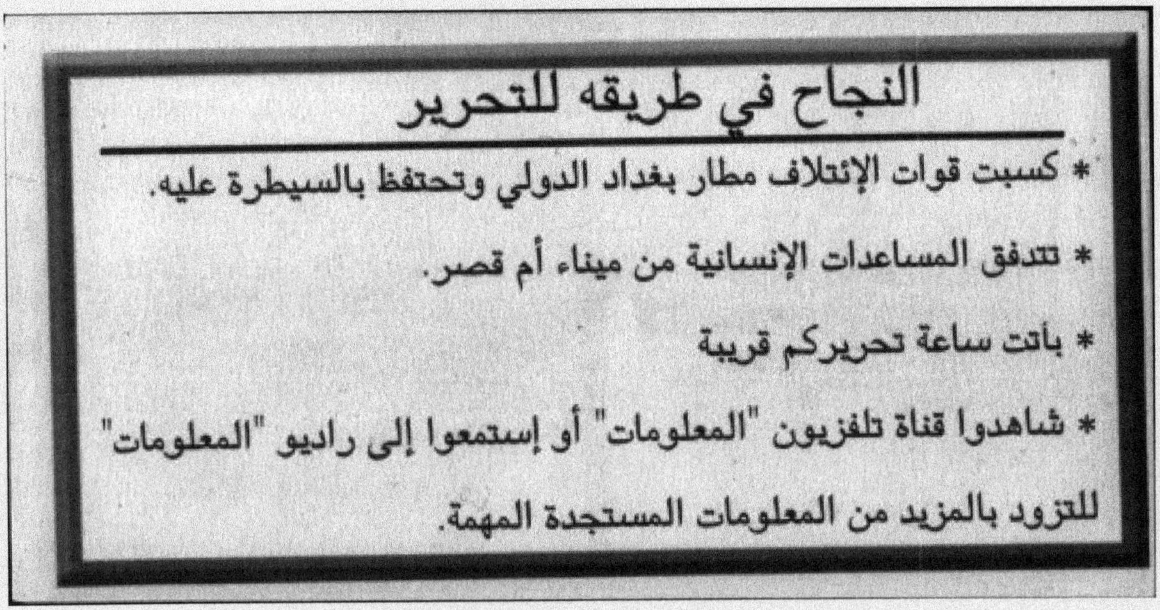

النجاح في طريقه للتحرير

* كسبت قوات الإئتلاف مطار بغداد الدولي وتحتفظ بالسيطرة عليه.

* تتدفق المساعدات الإنسانية من ميناء أم قصر.

* باتت ساعة تحريركم قريبة

* شاهدوا قناة تلفزيون "المعلومات" أو إستمعوا إلى راديو "المعلومات"

للتزود بالمزيد من المعلومات المستجدة المهمة.

Success is on the way to liberation.

- Coalition forces gain control of Baghdad International Airport.
- Humanitarian aid flows from Umm Qasr port.
- The hour of your liberation is near.
- Watch the "Information" TV channel or listen to "Information" Radio.
- To obtain more important new information.

Arabic

Leaflet 193

Nobody benefits from the use of weapons of mass destruction.

Arabic

Leaflet 194

Iraq.
We can see everything.
Do not use nuclear, biological, or chemical weapons.

Arabic

Leaflet 195

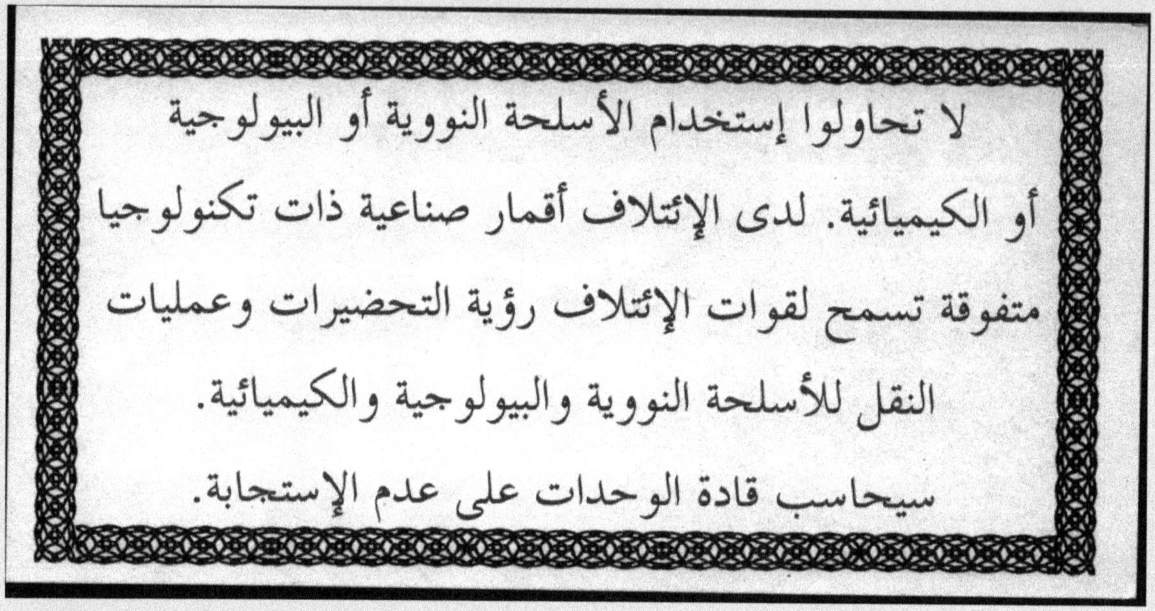
لا تحاولوا إستخدام الأسلحة النووية أو البيولوجية
أو الكيميائية. لدى الإئتلاف أقمار صناعية ذات تكنولوجيا
متفوقة تسمح لقوات الإئتلاف رؤية التحضيرات وعمليات
النقل للأسلحة النووية والبيولوجية والكيميائية.
سيحاسب قادة الوحدات على عدم الإستجابة.

Do not attempt to use nuclear, biological, or chemical weapons. The coalition has advanced industrial satellites with technology that allows them to see preparations and operations for the transportation of nuclear, biological, and chemical weapons. Unit commanders will be held accountable for non-compliance.

Arabic

Leaflet 196

Do not interfere with coalition forces. The coalition does not wish to harm the people of Iraq.

Arabic

Leaflet 197

The 1st Tammuz Battalion of the 24th Mechanized Brigade, 10th Armored Division did not comply with Coalition guidelines.

Arabic

Leaflet 198

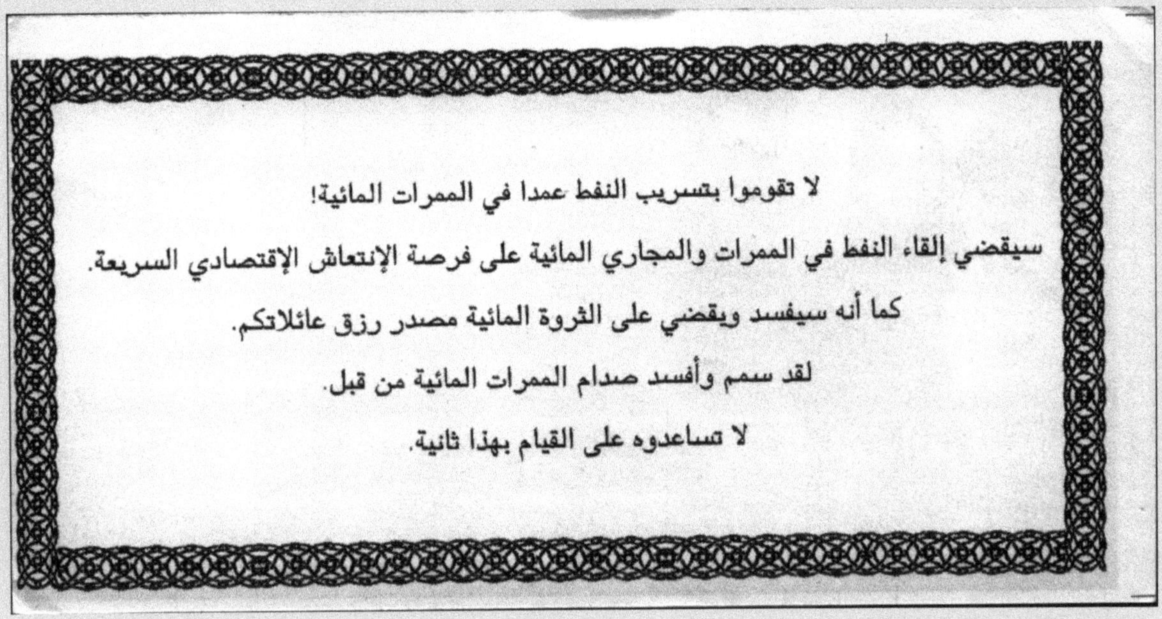
لا تقوموا بتسريب النفط عمدا في الممرات المائية!

سيقضي إلقاء النفط في الممرات والمجاري المائية على فرصة الإنتعاش الإقتصادي السريعة.

كما أنه سيفسد ويقضي على الثروة المائية مصدر رزق عائلاتكم.

لقد سمم وأفسد صدام الممرات المائية من قبل.

لا تساعدوه على القيام بهذا ثانية.

DO NOT RELEASE OIL INTO WATERWAYS!

Dumping oil in the waterways will ruin the chance for economic recovery. It will kill or taint the sea life, which feeds your family. Saddam has poisoned your waterways before.

You must not aid him in doing it again.

Arabic

Leaflet 199

Dumping oil poisons Iraqi waterways, as well as your family's future.

Arabic

Leaflet 200

Saddam's regime did not protect your family!
It was not just your family that they neglected.
It was all Iraqi families that they neglected!

Arabic

Leaflet 201

Return to your home!
Your families are waiting for you!

Arabic

Leaflet 202

For your safety do not interfere with Coalition Forces. Coalition Special Operation Forces are here to end the oppressive rule of Saddam's regime and liberate the people of Iraq.

Arabic

Leaflet 203

O Iraqi citizens! To ensure your safety, please do not carry weapons! You might be mistakenly identified as Saddam's soldiers or his security forces.

Arabic

Leaflet 204

The coalition forces do not want to harm the Iraqi people.

For your safety and the safety of others

Do not carry weapons.

Stay at home!

Arabic

Leaflet 205

If Saddam uses chemical weapons, the victims will be your comrades and the innocent Iraqi people.

You are accomplices to this crime.

Unit commanders will bear personal responsibility and be held accountable for their use of chemical weapons.

Arabic

Leaflet 206

Coalition forces are prepared and well trained to defend itself against chemical weapons attacks.

Arabic

Leaflet 207

Coalition forces are conducting operations in your areas. They are not here to harm you. Soon you will receive instructions regarding humanitarian aid.

Arabic

Leaflet 208

للحفاظ على سلامتكم

توقفوا عن إصلاح أسلاك الأ لياف الضوئية العسكرية.

أنتم تجازفون بحياتكم.

الأ سلاك هي أداة تستخدم من قبل صدام ونظامه في سبيل قمع الشعب العراقي ، فهذه
الأ سلاك مستهدفة لتدميرها.

To keep you safe

Stop repairing military fiber optic cords. You are risking your life. The wires are a tool used by Saddam and his regime to suppress the Iraqi people. These wires are targeted for destruction.

Arabic

Leaflet 209

Do not become a refugee! You and your family will face hardship and suffering, nothing else! For your safety, stay in your homes and villages.

Arabic

Leaflet 210

The coalition has amassed a massive fighting force in the north.

Arabic

Leaflet 211

Facing them (the coalition forces) will certainly lead to your death and the destruction of your family.

Arabic

Leaflet 212

This is what the leaflets say:

'You will be destroyed if you target coalition aircraft

.Arabic

Leaflet 213

If you ignore this warning, you will die!

Will this be your fate?

Don't target the coalition aircraft.

Arabic

Leaflet 214

I am sorry. I warned you not to target the coalition planes.

Arabic

Leaflet 215

The coalition aircraft targeted this unit.
I told you they wouldn't spare you if you didn't hide them.
Which fate do you choose?"

Arabic

Leaflet 216

Saddam loses the support of his military forces.

Arabic

Leaflet 217

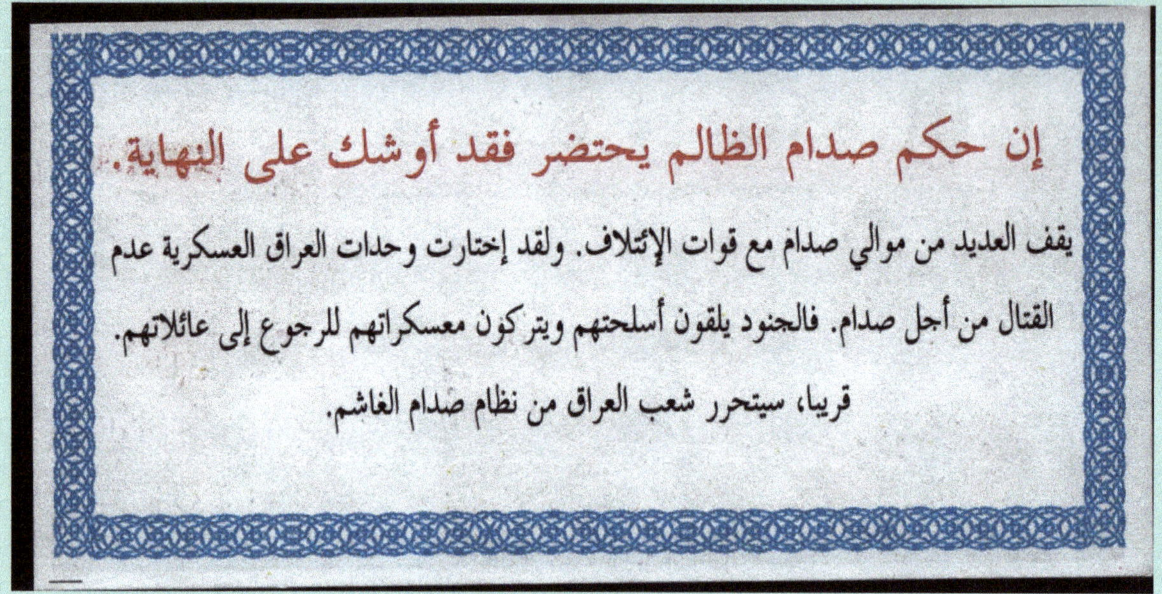

إن حكم صدام الظالم يحتضر فقد أوشك على النهاية.

يقف العديد من موالي صدام مع قوات الإئتلاف. ولقد إختارت وحدات العراق العسكرية عدم القتال من أجل صدام. فالجنود يلقون أسلحتهم ويتركون معسكراتهم للرجوع إلى عائلاتهم.

قريبا، سيتحرر شعب العراق من نظام صدام الغاشم.

Saddam's oppressive rule is dying. It is about to end. Many Saddam loyalists stand with the coalition forces. Iraq's military units chose not to fight for Saddam. The soldiers lay down their weapons and leave their camps to return to their families. Soon, the people of Iraq will be liberated from Saddam's brutal regime.

Arabic

Leaflet 218

To maintain your safety, respond to the coalition's directives.

Arabic

Leaflet 219

للحفاظ على سلامتكم

أظهروا أنكم لا تشكلون تهديدا وذلك بإتباع توجيهات الإئتلاف. إبتعدوا مسافة آمنة عن
المضادات للطائرات والمدفعية ولكن لا تتركوا منطقة القاعدة. إرموا أسلحتكم ولا تتدخلوا
بعمليات الإئتلاف.

To keep you safe

Show that you are not a threat by following coalition directions. Stay a safe distance away from anti-aircraft and artillery, but do not leave the base area. Lay down your weapons and do not interfere with coalition operations.

Arabic

Leaflet 220

Take an aggressive position and you will be destroyed.

Arabic

Leaflet 221

Do not take an aggressive position, you will not be destroyed.

Arabic

Leaflet 222

Save your life by following the coalition's directives.

Park all aircraft on the airfields with their noses facing each other.

Do not block the runways.

Abandon anti-aircraft equipment and artillery; they will be destroyed on sight.

Arabic

Leaflet 223

In the event of the outbreak of fighting, Iraqi units that do not surrender immediately will be destroyed by coalition ships and aircraft.

Arabic

Leaflet 224

The July 1st Tank Battalion of the 24th Mechanized Brigade in the 10th Armored Division did not respond to coalition instructions.

Arabic

Leaflet 225

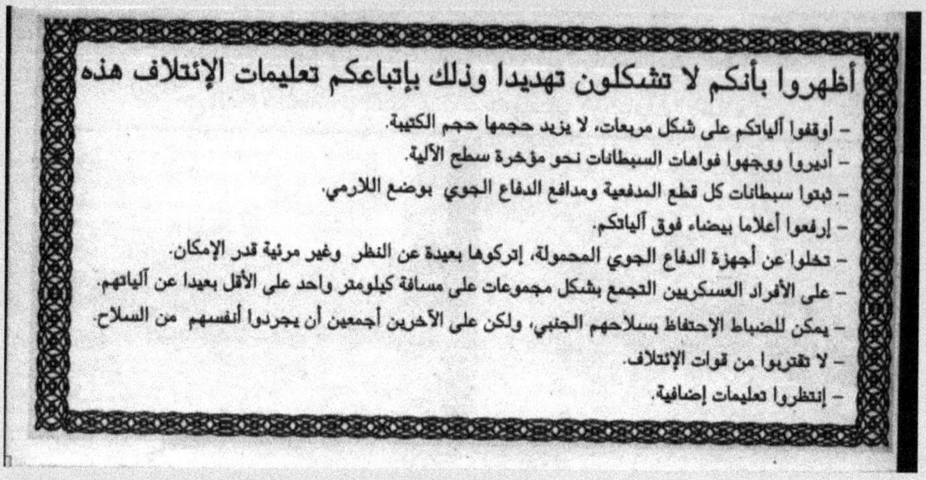
أظهروا بأنكم لا تشكلون تهديدا وذلك بإتباعكم تعليمات الإئتلاف هذه

- أوقفوا آلياتكم على شكل مربعات، لا يزيد حجمها حجم الكتيبة.
- أديروا ووجهوا فواهات السبطانات نحو مؤخرة سطح الآلية.
- ثبتوا سبطانات كل قطع المدفعية ومدافع الدفاع الجوي بوضع اللارمي.
- إرفعوا أعلاما بيضاء فوق آلياتكم.
- تخلوا عن أجهزة الدفاع الجوي المحمولة، إتركوها بعيدة عن النظر وغير مرئية قدر الإمكان.
- على الأفراد العسكريين التجمع بشكل مجموعات على مسافة كيلومتر واحد على الأقل بعيدا عن آلياتهم.
- يمكن للضباط الإحتفاظ بسلاحهم الجنبي، ولكن على الآخرين أجمعين أن يجردوا أنفسهم من السلاح.
- لا تقتربوا من قوات الإئتلاف.
- إنتظروا تعليمات إضافية.

Show that you are not a threat by following these coalition instructions:

- Park your vehicles in squares, no larger than the size of the battalion.
- Rotate and direct the barrel nozzles toward the back of the mechanism deck.
- Install the barrels of all artillery pieces and air defense guns in the firing position.
- Raise white flags over your vehicles.
- Abandon portable air defense devices, leave them out of sight and invisible as much as possible.
- Military personnel must gather in groups at a distance of at least one kilometer away from their vehicles.
- Officers can keep their sidearm, but all others must disarm themselves.
- Do not approach coalition forces.
- Wait for additional instructions.

Arabic

Leaflet 226

Attention, Special Republican Guard soldiers, the coalition forces are coming to rid Iraq of Saddam and his regime. You must not share their fate.

Arabic

Leaflet 227

Do not interfere with the coalition forces, stay in your homes.

For your protection, coalition forces will impose a curfew.

Tune your radios to the "Information" radio waves to receive the latest developments.

Arabic

Leaflet 228

The coalition does not want to destroy your country's landmarks. The coalition will destroy any vital military targets.

Arabic

Leaflet 229

The proud Iraqi people are not the target of coalition military operations.
A strict and violent regime is the goal of coalition military operations.

Arabic

Leaflet 230

How much more?

Your families or the regime.

Arabic

Leaflet 231

While your families struggle to stay alive.
He lives in luxury and luxury.

Arabic

Leaflet 232

Leave now and return to your villages.

Take care of your children, nurture them, educate them, and ensure their success.

Arabic

Leaflet 233

Don't risk your life.
And the lives of your comrades.

Arabic

Leaflet 234

Think!

Is this the future you seek?

Your family needs you most!

Arabic

Leaflet 235

Join your family.
Leave the battlefield.
And leave it behind.
Save yourself.

Arabic

Leaflet 236

To maintain your safety, stay away from military operations near waterways.

Arabic

Leaflet 237

Dumping oil into waterways will not hinder coalition operations.

Arabic

Leaflet 238

Intentionally leaking oil into waterways destroys water resources and destroys Iraq's economic growth opportunities.

Arabic

Leaflet 239

Coalition forces do not wish to harm innocent civilians.
Once Saddam's regime is defeated, the waterways will be returned to the Iraqi people.

Arabic

Leaflet 240

Do not assist Iraqi military personnel or regime leaders who are trying to escape.

Arabic

Leaflet 241

إنتباه

إن قوات الإئتلاف موجودة هنا للتصدي للقادة العراقيين الذين يحاولون الهرب.

إذا قمتم بإيواء أو مساعدة أو إعانة هؤلاء الأشخاص فإنكم تعرضون أنفسكم وطاقمكم

للخطر كما أنكم تجازفون بمصادرة مراكبكم .

إن لاحظتم أن هناك منشقين فارين بلغوا قوات الإئتلاف عنهم.

لا تدعوا نظام صدام يفلت من العدالة.

Attention

Coalition forces are here to confront Iraqi leaders who are trying to escape. If you shelter, assist, or aid these people, you are putting yourself and your crew at risk, and you are also risking the confiscation of your boats. If you notice that there are defectors fleeing report them to coalition forces. Do not let Saddam's regime escape justice.

Arabic

Leaflet 242

Your choice

Arabic

Leaflet 243

If
Then

Arabic

Leaflet 244

Military fiber optic cords

Arabic

Leaflet 245

Military fiber optic wires were targeted and destroyed. Repairing these wires puts your lives in danger.

Arabic

Leaflet 246

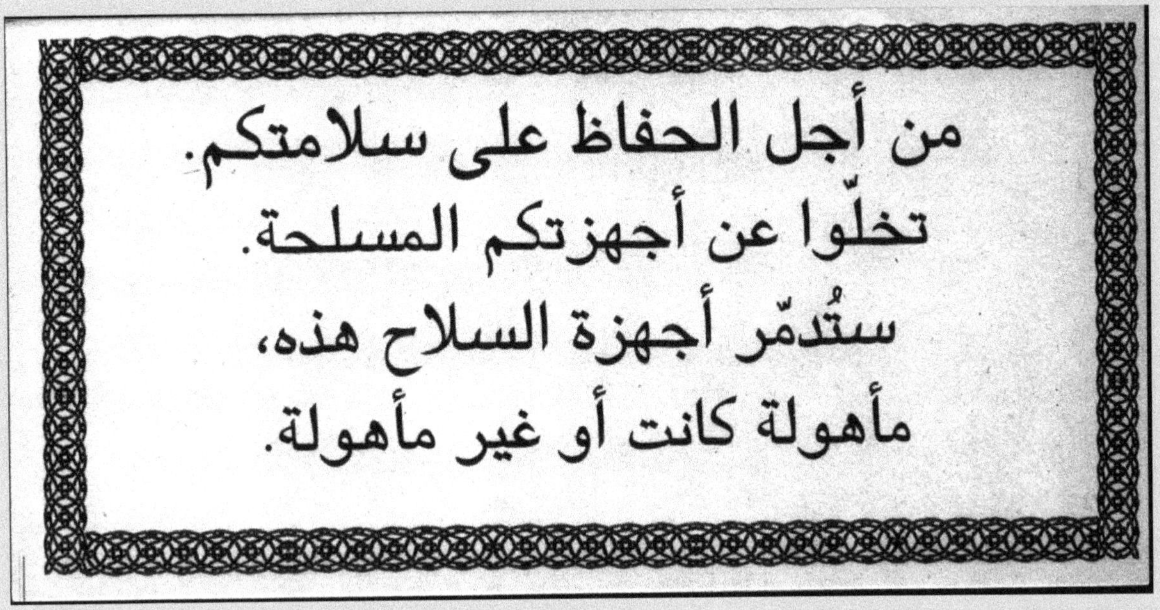

من أجل الحفاظ على سلامتكم.
تخلّوا عن أجهزتكم المسلحة.
ستُدمّر أجهزة السلاح هذه،
مأهولة كانت أو غير مأهولة.

In order to keep you safe. Abandon your armed devices. These weapons will destroy weapons, whether manned or unmanned.

Arabic

Leaflet 247

Air defense artillery units have been targeted and destroyed.

Arabic

Leaflet 248

Mining waterways will not affect coalition ships. Any ship suspected of mining waterways will be destroyed.

Arabic

Leaflet 249

Inform coalition forces of mining operations. Mined waterways will only harm the Iraqi people.

Arabic

Leaflet 250

Mining the Khar Abdullah or Umm Qasr waterways would affect coalition ships. Mining will only harm the Iraqi people.

Arabic

Leaflet 251

Any ship that is suspected of intending to block the passages of 'Khur Abdullah' or 'Umm Qasr' will be destroyed.

Arabic

Leaflet 252

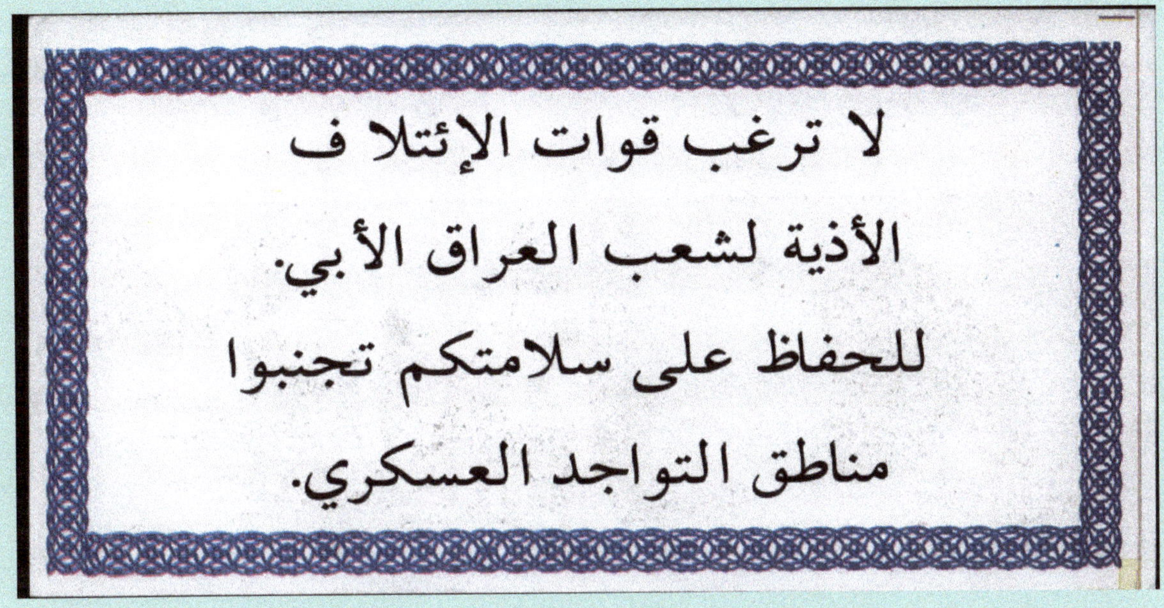

The coalition forces do not wish to harm the proud people of Iraq. To maintain your safety, avoid military presence areas.

Arabic

Leaflet 253

If you move from your current position, you will be attacked by the coalition forces. For your safety, remain in your positions and wait for the coalition's instructions.

The coalition is here to end the unjust rule that Saddam and his regime have established.

Arabic

Leaflet 254

Stay where you are.

Remain in your places and await the coalition's instructions. Any movement from your current locations will be considered a hostile act. This may lead to you being attacked by coalition forces.

Arabic

Leaflet 255

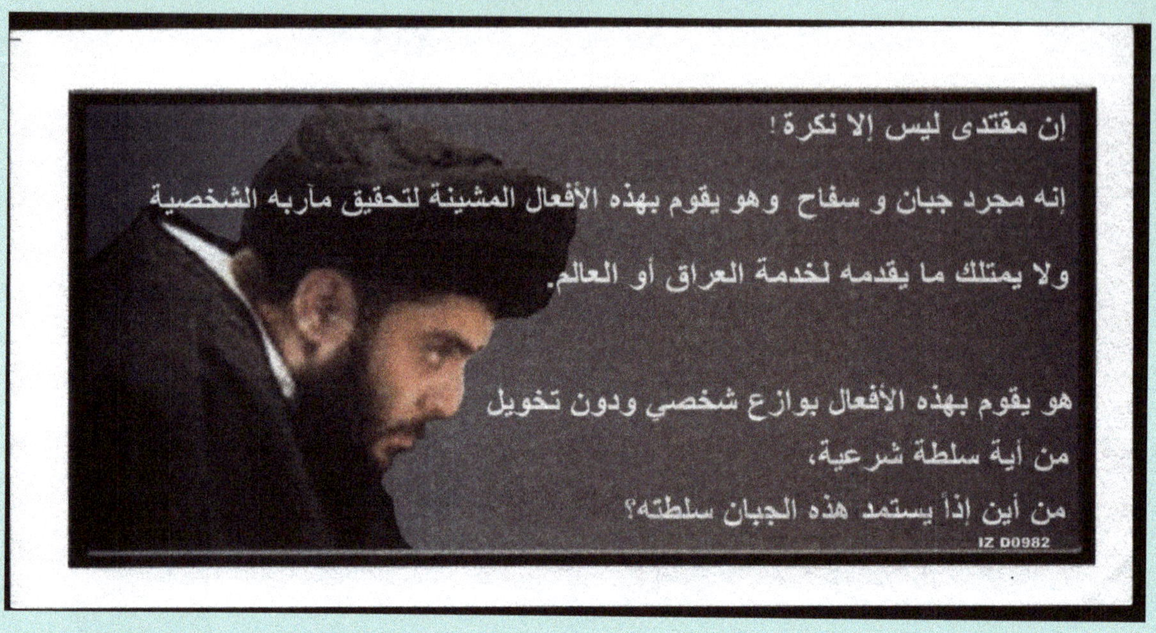

Muqteda is nothing but a nobody!

He is merely a coward and a butcher, carrying out despicable acts to fulfill his personal ambitions. He has nothing to offer for the service of Iraq or the world.

He carries out these actions with personal motives and without any legitimate authority.

Where does this coward derive his authority from?

Arabic

Leaflet 256

What the criminal Muqtada is doing is desecrating the holy places and exploiting them as hideouts for weapons and terrorists.

He has no respect for the sanctity of these sites or the people of Najaf.

Return the shrine of Imam Ali to the seminary.

Arabic

Leaflet 257

Throwing stones is dangerous!

When a child throws stones at coalition vehicles, this may be misunderstood by the soldiers as an offensive act that results in them being injured or killed. For the safety of everyone, no one should throw stones at vehicles and transporters of the coalition forces. No one should throw stones at the coalition forces for the safety of the soldiers.

Coalition forces are here to protect you

Arabic

Leaflet 258

Throwing rocks is dangerous.

It may be fun to throw rocks at Coalition Forces' vehicles, but this could cause injury to coalition soldiers or injuries to you.

Coalition Forces are authorized to arrest any person throwing rocks or participating in any other dangerous activities.

Arabic

Leaflet 259

Real Iraqi Heroes.

Arabic

Leaflet 260

Good Citizens of Mosul

Thank you for your support and cooperation during these trying times for our city. We are working hard to make your homes safe once again.

Arabic

Leaflet 261

Ansar Al-Sunna controls Al-Qaeda.

Al-Qaeda.

They are both takfiris.

Arabic

Leaflet 262

Al-Qaeda controls Ansar Al-Sunnah
Al-Qaeda
Ansar Al-Sunnah
They are both takfiris.

Arabic

Leaflet 263

Your attack on coalition aircraft will lead to your destruction.

Arabic

Leaflet 264

Do not shoot at the coalition plane. If you choose to fire, you will be destroyed. Coalition forces will attack with overwhelming force. You have the choice.

Arabic

Leaflet 265

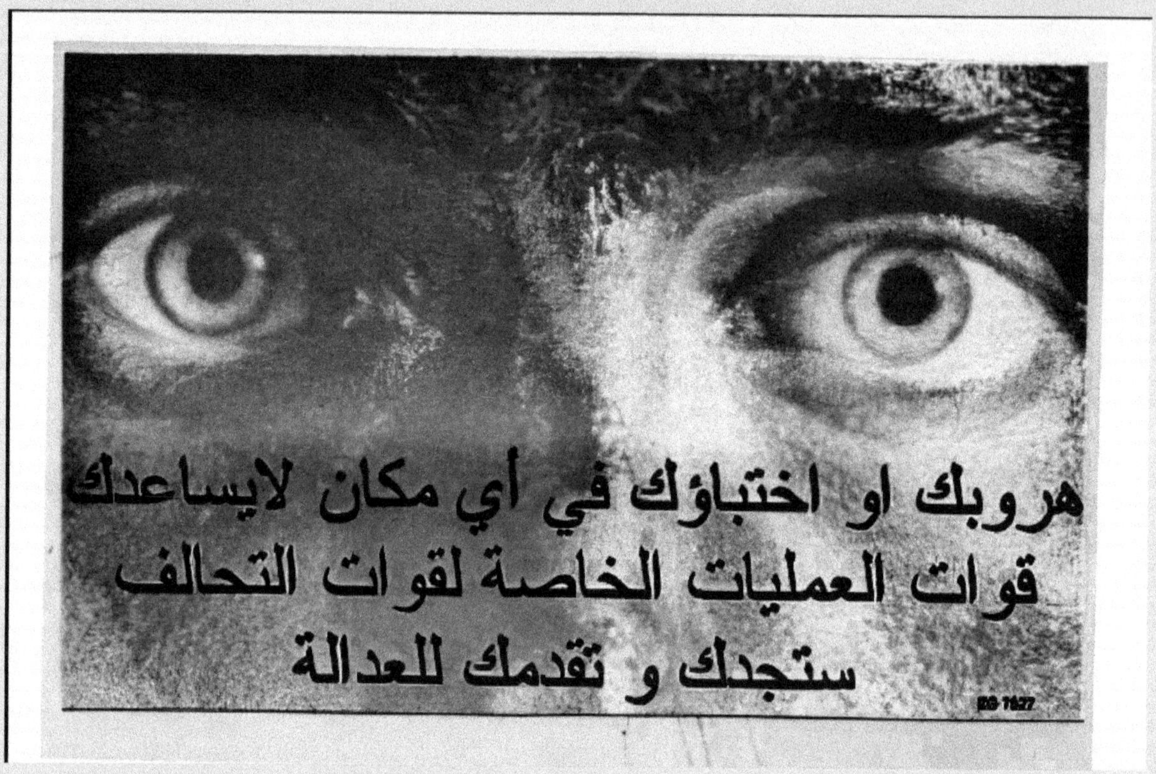

No matter where you run, no matter where you hide, Coalition Special Operations Forces will find you and bring you to justice.

Arabic

Leaflet 266

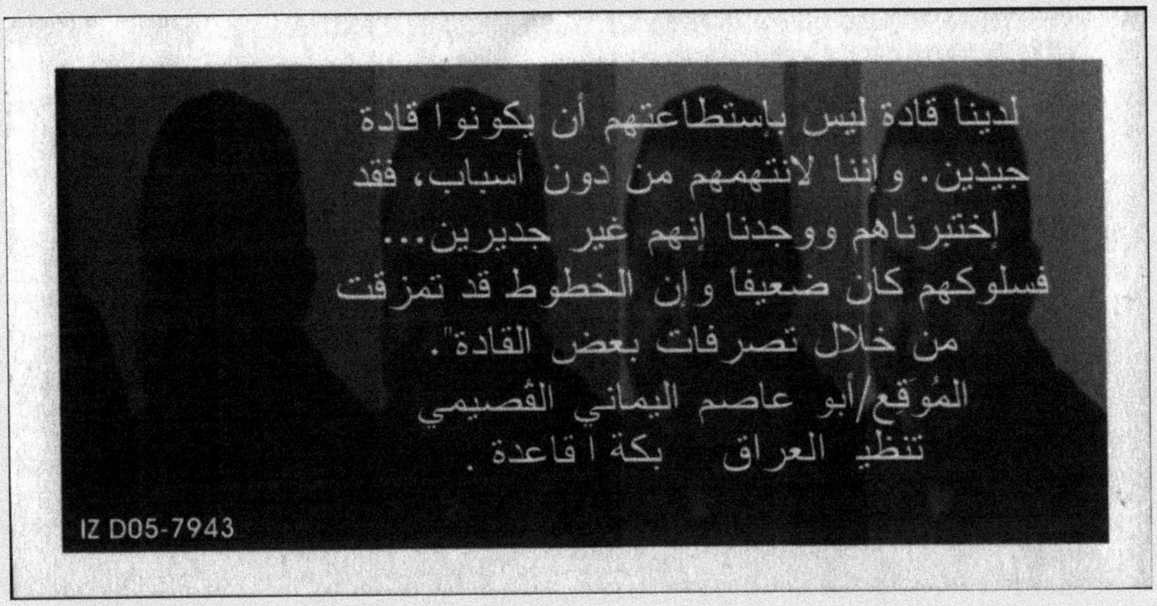

We have leaders who are unable to be good leaders. We do not blame them without reason; we tested them and found them to be unworthy. Their behavior was weak, and the lines may have been crossed due to the actions of some leaders."

Signed: Abu Asim Al-Yamani Al-Qusaymi Iraq's Al-Qaeda Leadership.

Arabic

Leaflet 267

We are coming to destroy you!!

We will kill or capture all members of the Al-Qaeda organization.

There are those among you who have betrayed you and provided us with the information about your whereabouts, no matter where you hide!

Turki Hisham Turki Hisham - Al-Qaeda Organization: Rabat Organization

Arabic

Leaflet 268

Make your city a place of peace and comfort as much as you can.

Report terrorist activities.

Please call the following numbers:

AsiaCell: 0770-0394497, 0770-0445070

Iraqna: 0790-1376347, 0790-1380248

Satellite Line: 88216-3124813157, 88216-3124813158

Email: tipstallafar@yahoo.com

You can remain anonymous, but be assured that any information you provide will be treated with the utmost confidentiality.

You will receive a financial reward, and your city will once again become a great place to live.

Arabic

Leaflet 269

The Iraqi security forces are fighting hard to make our country a safe and secure city. We thank you for your cooperation and support for the Iraqi security forces, who have brought terrorists to justice.

Arabic

Leaflet 270

Kill

For their crimes against the Iraqi people

Mohammad Jassem and Omar

Whoever harbors them is considered a murderer and an enemy of the people of Al-Siniyah. They will not only harm you and your children but also destroy your families. So act now and support the Iraqi security forces by contacting the tip line with any information about their terrorist activities. Help bring these terrorists to justice.

Anonymous tip lines: (0770) 614-2438 (0770) 282-0464

Arabic

Leaflet 271

مطلوب
هؤلاء مشتبه بهم بقيام هجمات التي قتلت العزل من
العراقيين لديك القوة لايقاف الاعداء من التقدم و الحرية.
احسان محسن كريم الحداد
سليمان داوود بكر
الله أكبر
اي شخص مؤهل للحصول على مكافاة اذا وفر اي معلومات
تقود الى اعتقال هؤلاء المجرمين.

Wanted

These individuals are wanted for conducting attacks that have resulted in the deaths of innocent Iraqis. Your strength lies in stopping these enemies of progress and freedom.

Ahmad Mahmoud Talib Al-Faraji

Abu Mus'ab Al-Zarqawi (36,000,000 Iraqi Dinars bounty)

Hassan Mohsen Kareem Al-Haddad

Sami Mustafa Farhan Al-Hijazi

Haydar Ali Hussein Hasan Al-Daraji

Saif Mohammad Zayer Al-Mashhadani

Mahmoud Mohammad Sikka

If you have any credible information leading to the arrest of these criminals, report it to the local Iraqi security forces.

Arabic

Leaflet 272

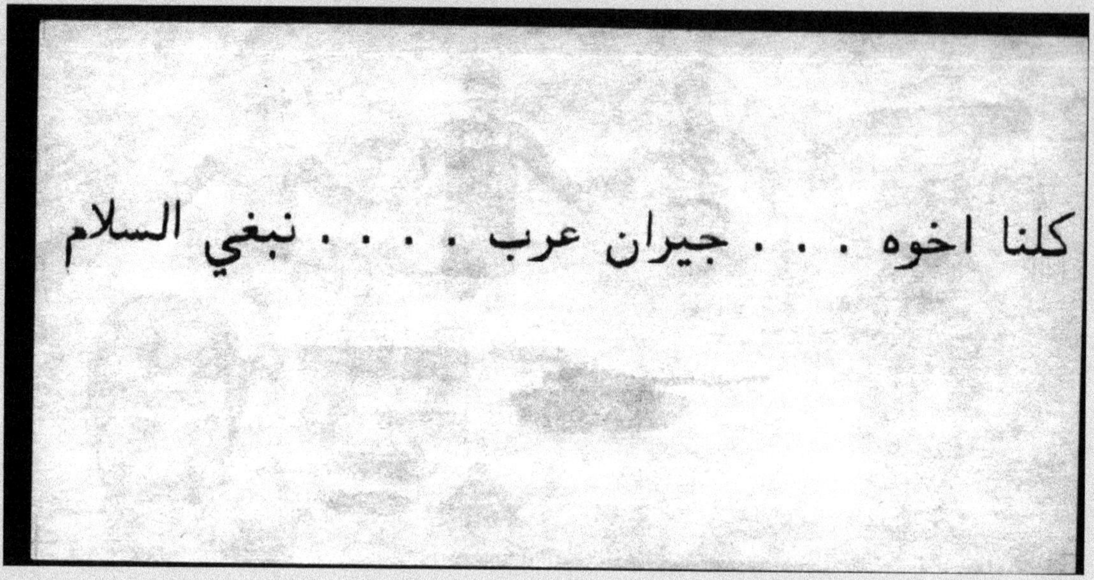

We are all brothers . . . Arab neighbors. . . . We want peace.

Arabic

Leaflet 273

Coalition forces are fully committed to the liberation and the good of the Iraqi people.

Your contribution to this effort will accelerate the flow of humanitarian aid and assistance to the Iraqi people.

Arabic

Leaflet 274

The Coalition forces will liberate Iraqi citizens.

Do your best to help liberate Iraq.

Do not steal or do revengeful or violent acts Put your disagreements aside until Iraq is stable and until Iraq is rebuilt again.

Arabic

Leaflet 275

إعلان Notice

المعتقلون الذين رغم معتقداتهم الدينية أو السياسية ، يرون أن حياتهم في خطر ، أو أنهم سوف يتعرضون لإصابات جسدية على أيدي الأسرى الآخرين عليهم القيام بالتبليغ عن هذه الحالة شخصياً إلى أي ضابط في الجيش الأمريكي في هذا المعسكر دون استشارة مندوب المعتقلين . ومن ذلك الوقت فصاعداً، سوف يقوم قائد المعسكر بتأمين الحماية الكافية لهؤلاء المعتقلين ، إما بالفصل بينهم وبين الآخرين أو نقلهم أو أي وسائل أخرى . وسوف يعاقب المعتقلون الذين يسيئون معاملة زملائهم المعتقلين .

القائد

Notice

Detainees who, despite their religious or political beliefs, see their lives in danger, or believe they will be subjected to physical harm by other detainees, should immediately report this to any U.S. Army officer in this camp without consulting the detainees' representative. From that moment on, the camp commander will ensure these detainees are adequately protected, either by separating them from others or by other means. Detainees who mistreat their fellow detainees will be punished.

The Commander

Arabic

Leaflet 276

FOR YOUR SAFETY

Stay in your homes and avoid driving your vehicles at night.

Stay away from buildings that have military weapons near them.

FOR YOUR SAFETY.

Arabic

Leaflet 277

Daily
from 6 pm until 11 pm
Information radio
On the following waves
Medium 756 kHz AM.
Medium 190 kHz AM.
Short 9715 kHz SW
Short 11292 kHz SW
And wave 4 is 100 MHz EM
Daily
from 6 pm until 11 pm

Arabic

Leaflet 278

Information radio
Daily from 6 pm until 11 pm
Medium 756 kHz AM.
Medium 693 kHz AM.
Short 9715 kHz SW
Short 11292 kHz SW
And the wave is 4,100 MHz FM

Arabic

Leaflet 279

Information for Liberation

O Iraqi citizens, receive important information from the coalition forces daily from 6:00 PM to 11:00 PM on Channel 1."

Adjust your radio devices to the following frequencies for information:

Medium Wave 756 kHz from 6:00 PM to 11:00 PM

Medium Wave 920 kHz from 6:00 PM to 11:00 PM

Medium Wave 1000 kHz from 6:00 PM to 11:00 PM"

Arabic

Leaflet 280

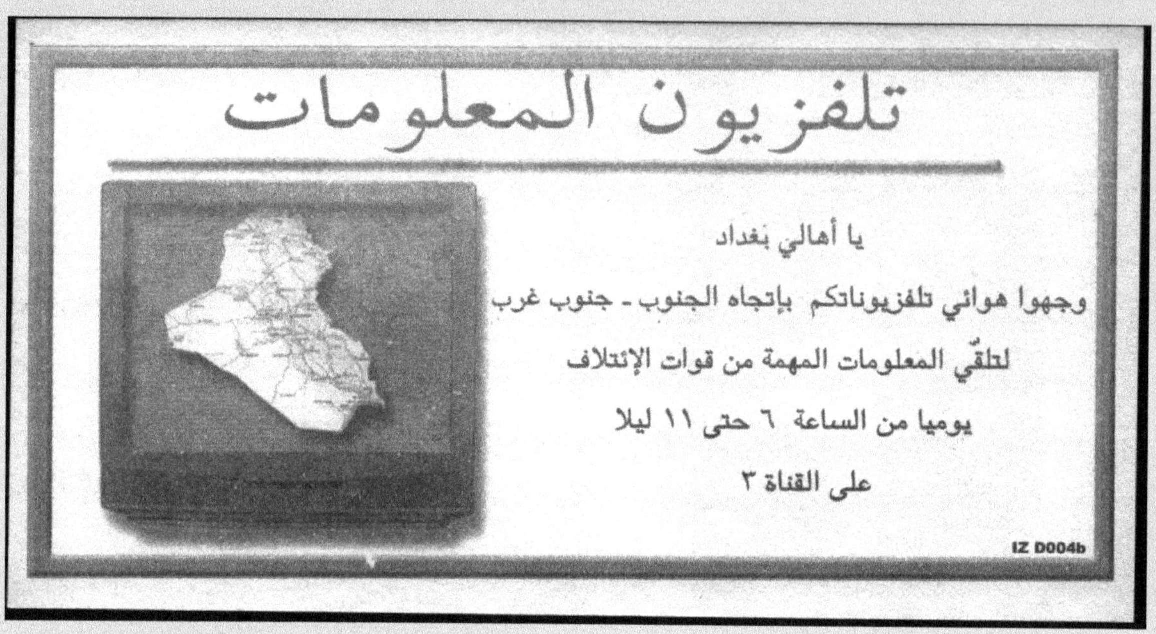

Information Television

Families of Baghdad.

Turn your television antennas to face SSW.

To receive important information from the Coalition Forces.

Daily from 6 am until 11 pm.

On Channel 3.

Arabic

Leaflet 281

Information radio
Daily
From 6 pm
Until 11 at night
Medium 756 kHz AM.
Medium 693 kHz AM.
Short 9715 kHz SW
Short 11292 kHz SW
And the wave is 4,100 MHz FM

Arabic

Leaflet 282

Stay at home with your families in safety and peace.

Arabic

Leaflet 283

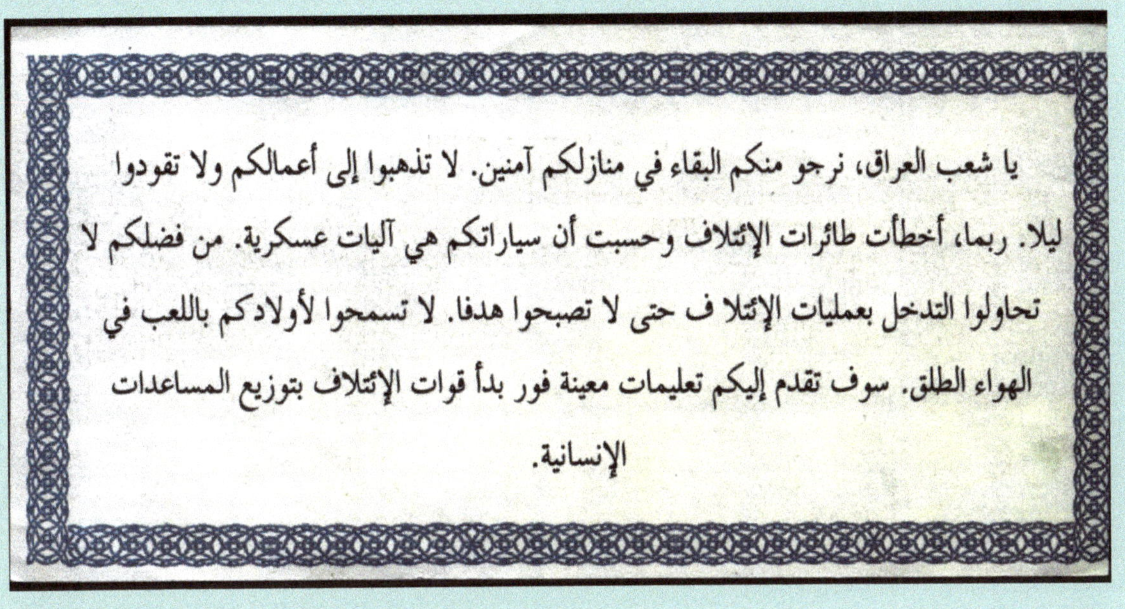
يا شعب العراق، نرجو منكم البقاء في منازلكم آمنين. لا تذهبوا إلى أعمالكم ولا تقودوا
ليلا. ربما، أخطأت طائرات الإئتلاف وحسبت أن سياراتكم هي آليات عسكرية. من فضلكم لا
تحاولوا التدخل بعمليات الإئتلا ف حتى لا تصبحوا هدفا. لا تسمحوا لأولادكم باللعب في
الهواء الطلق. سوف تقدم إليكم تعليمات معينة فور بدأ قوات الإئتلاف بتوزيع المساعدات
الإنسانية.

People of Iraq, please stay home and stay safe. Do not go to work and do not drive at night. Perhaps, the coalition aircraft mistook your vehicles for military vehicles. Please do not try to interfere with the coalition's operations so that you do not become a target. Do not allow your children to play outdoors. Specific instructions will be provided to you once Coalition Forces begin distributing aid Humanity.

Arabic

Operation Inherent Resolve

(June 15, 2014 – Ongoing)

Leaflet 284

When you hear an approaching plane!

- Lay on your side.
- Cover your ears.
- Close your eyes.
- Open your mouth.
- Keep your chest from touching the ground.

Coalition forces to not target civilians.

Arabic

Leaflet 285

Keep away from Da’esh as much as possible.

The Coalition is destroying Da’esh buildings while seeking civilians' safety.

Arabic

Leaflet 286

ISIS robs our boys of their childhood… and turns them into murderers.

Arabic

Leaflet 287

Coalition pilots prowl the skies. Listen to their roar and watch for the fire and smoke that signifies that another ISIS fighter has been sent to hell.

Arabic

Leaflet 288

Daesh.

Arabic

Leaflet 289

Allah u Akbar.

ISIS.

Fallujah.

Arabic

Leaflet 290

The Iraqi Security Force is advancing…Quick, just like we did at Fallujah.

Arabic

Iraq

Operation Inherent Resolve - 2014

Leaflet 291

Leaflet 292

For any oppressor there is an end.

Arabic

Leaflet 293

Arabic

Leaflet 294

ISIS uses innocent civilians as human shields.
They force them to be in the places which are targeted for air strikes.

Arabic

Leaflet 295

Attention Syrian People: Go away from Daesh! The Coalition will be dropping bombs on the ISIS fighters.

Arabic

Leaflet 296

Leave the fight now. Do not take yourself into destruction.

Arabic

Leaflet 297

Don't let Daesh take you with them to their inevitable fate.

Arabic

Leaflet 298

The time you have been waiting for has come, the time to leave Raqqa.

Arabic

Leaflet 299

To those involved in carrying weapons, we stretch out our hand to you. Reserve your place before it is too late.

Arabic

Leaflet 300

Waving a white flag to Syrian Democratic Forces is a sign from you that you do not want to fight. Follow the instructions of Syrian democratic solders and they will treat you well and fairly once you surrender.

Arabic

Leaflet 301

ISIS is using boats and ferries to transport weapons and fighters. Do not use ferries or boats, air strikes are coming.

Arabic

Leaflet 302

Departure Instructions

Keep this secret and keep this ticket until you reach a safe place.

Do not tell anybody of your departure.

Head toward the Syrian democratic forces peacefully and without weapons.

If possible, wave a piece of white cloth to surrender.

Follow all the steps to guarantee your safety and you will be treated humanely and with respect.

Freeing Raqqa has started and if you stay it will lead to your death.

ISIS will fall and you should not be there when that happens.

Arabic

Leaflet 303

Daesh will lose the city of Tabqa, just as it lost Manbij and Raqqa.

Arabic

Leaflet 304

The Civil Council of Tabqa and the Internal Security Forces in Raqqa are working to remove all unexploded materials and devices. Fill out this form and submit it to the Internal Security Forces checkpoints in Raqqa to remove these materials.

Arabic

Leaflet 305

Stay away from these locations during coalition strikes.

Fighting positions for Da'esh.

Da'esh check points.

Coalition forces realize you are victims of Da'esh terror by using you as human shields.

Arabic

THE LIBYAN CIVIL WAR

Operation Odyssey Dawn

(March 19, 2011 – March 31, 2011)

Leaflet 306

Warning
Step away from military activities.

Arabic

Leaflet 307

NATO is here to protect civilians.

Arabic

Leaflet 308

English

Leaflet 309

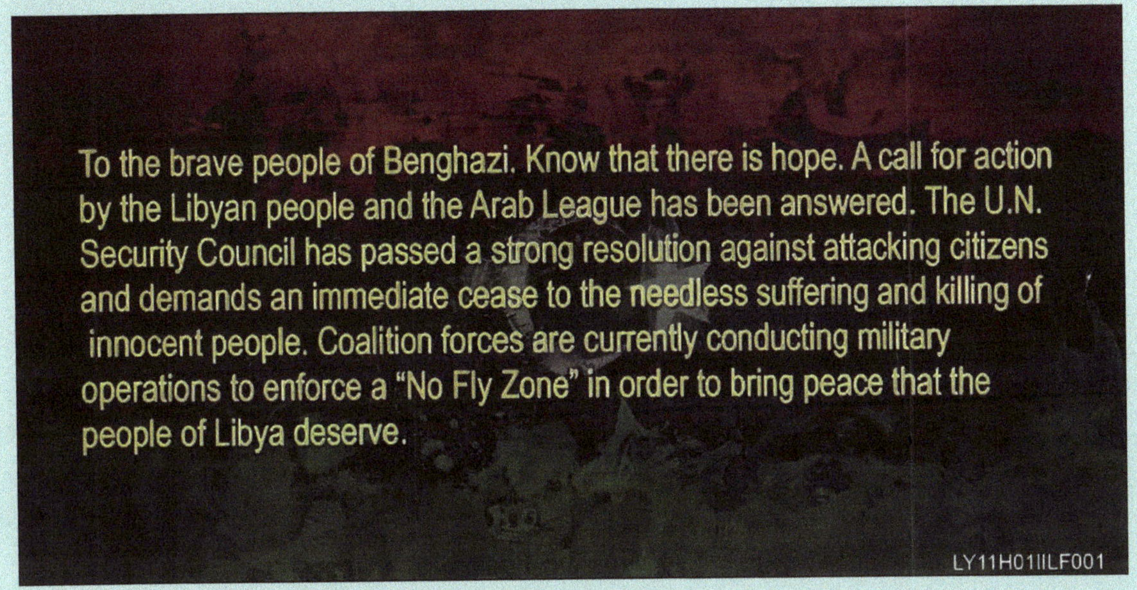

English

Leaflet 310

NATO air forces have imposed a no-fly zone for the sake of protecting Libyan civilians.

Arabic

Leaflet 311

Stop tearing Libya apart.
One Libya, one people.

Arabic

Leaflet 312

Soldiers committing crimes against humanity will be held accountable by the international community. Gaddafi has been indicted by the International Criminal Court. Will you share a prison cell with him? Who will support your family? Make a choice before it is too late – peace and future prosperity or continued war and further death and destruction.

Arabic

Leaflet 313

Libya is one and its people are one, turn the dial to station 104.1 FM.

Arabic

Leaflet 314

Freedom of information is the foundation of the people's freedom.
Demand your legal right to obtain information for free.
Tune your radio to 104.1 FM and listen.

Arabic

Leaflet 315

To the leaders of the former Gaddafi regime in Sirte: Stop fighting!

Warning: Leave now!

Arabic

Leaflet 316

Reconciliation. Justice. Freedom. Stability. The future. Peace.

Arabic

Leaflet 317

Libya is one and its people are one.

Arabic

GLOSSARY

Title	Definitions
10th Armored Division	The 10th Armored Division was an armored division of the United States Army in World War II. In the European Theater of Operations, the 10th Armored Division was part of both the Twelfth United States Army Group and Sixth United States Army Group.
Abu Musab Al-Zarqawi	Abu Musab Al-Zarqawi, born Ahmad Fadeel Al-Nazal Al-Khalayleh, was a Jordanian jihadist who ran a terrorist training camp in Afghanistan.
Afghanistan	Afghanistan, officially the Islamic Emirate of Afghanistan, is a landlocked country located at the crossroads of Central Asia and South Asia.
Afghans	Afghans are people from Afghanistan, a country in South Asia, characterized by diverse ethnic groups including Pashtuns, Tajiks, Hazaras, Uzbeks, and others. They speak Dari and Pashto and have a rich cultural heritage shaped by their historical location and tribal traditions.
Al-Qaeda	Al-Qaeda was established from a network of Arab and other foreign veterans of the U.S.- backed Afghan insurgency against the Soviet Union, with the aim of supporting Islamist causes in conflicts around the world.

Ayman Al-Zawahiri	Ayman Mohammed Rabie Al-Zawahiri was an Egyptian-born militant and physician who served as the second general emir of Al-Qaeda from June 2011 until his death in July 2022.
Baghdad	Baghdad is the capital of Iraq and the second-largest city in the Arab world after Cairo.
Bakkah	Bakkah is an ancient name for Mecca, the most holy city of Islam.
Bali	Bali is a province of Indonesia.
Basra	Basra is a city in southern Iraq.
CIA	The Central Intelligence Agency (CIA) is a civilian foreign intelligence service of the federal government of the United States
Coalition forces	The Multi-National Force – Iraq, often referred to as the Coalition forces, was a military command during the 2003 invasion of Iraq and much of the ensuing Iraq War, led by the United States of America, United Kingdom, Australia, Italy, Spain, and Poland, responsible for conducting and handling military operations.
Cold War	After World War II, the United States and its allies, and the Soviet Union and its satellite states began a decades-long struggle for supremacy known as the Cold War. Soldiers of the Soviet Union and the United States did not do battle directly during the Cold War. But the two superpowers continually antagonized each other through political maneuvering, military

	coalitions, espionage, propaganda, arms buildups, economic aid, and proxy wars between other nations.
Commonwealth Nations	The Commonwealth is a voluntary association of 56 independent and equal countries.
Daesh	Islamic State of Iraq and Syria (ISIS) and by its Arabic acronym Daesh, is a transnational Salafi jihadist group and a former unrecognized quasi-state.
Darri	Dari is the Afghan dialect of Persian and as such is sometimes referred to as "Eastern Persian". Dari, rather than Pashto, serves as the lingua franca for the languages of Afghanistan. Roughly one-third of the population of Afghanistan speak the language.
FM	FM broadcasting is a method of radio broadcasting that uses frequency modulation (FM) of the radio broadcast.
Geneva Convention	The Geneva Convention set out how soldiers and civilians should be treated in war.
Gulbuddin Hizb	The Hezb-e-Islami Gulbuddin also referred to as Hezb-e-Islami or Hezb-i-Islami Afghanistan (HIA), is an Afghan political party and paramilitary organization, originally founded in 1976 as Hezb-e-Islami and led by Gulbuddin Hekmatyar.
Gulf War	The Gulf War was an armed conflict between Iraq and a 42-country coalition led by the United States. The coalition's efforts against Iraq were carried out in two key phases: Operation Desert Shield, which marked the military buildup from August 1990 to January 1991; and Operation Desert Storm, which began with the aerial bombing campaign against Iraq on 17 January 1991 and came to a close with the American-led liberation of Kuwait on 28 February 1991.

Hamid Karzai	Hamid Karzai is an Afghan politician who served as the fourth president of Afghanistan.
Infantry Division	An infantry division consists chiefly of foot soldiers equipped with light weapons but also includes supporting artillery, armor, and engineer units and has its own communication, supply, maintenance, and evacuation services.
International Criminal Court (ICC)	It is the first and only permanent international court with jurisdiction to prosecute individuals for the international crimes of genocide, crimes against humanity, war crimes and the crime of aggression.
ISI	The Inter-Services Intelligence (ISI) is the largest and best-known component of the Pakistani intelligence community. It is responsible for gathering, processing, and analyzing any information from around the world that is deemed relevant to Pakistan's national security.
ISIS	Islamic State of Iraq and Syria (ISIS) and by its Arabic acronym Daesh, is a transnational Salafi jihadist group and a former unrecognized quasi-state.
Libya	It is the fourth-largest country in Africa and the Arab world, and the 16th-largest in the world.
Mine	A stationary explosive device that is designed to destroy personnel, ships, or vehicles when the latter encounter it.
Nairobi	Nairobi is the capital and largest city of Kenya.

NATO	The North Atlantic Treaty Organization (NATO) was founded in 1949 and is a group of 32 countries from Europe and North America that exists to protect the people and territory of its members.
No-Fly Zone Warning	Iraqi aircraft were forbidden from flying inside the zones. The policy was enforced by the United States and the United Kingdom until 2003, when it was rendered obsolete by the 2003 invasion of Iraq.
Operation Desert Shield	Operation was codenamed as Operation Desert Shield (2 August 1990 – 17 January 1991) for operations leading to the buildup of troops and defense of Saudi Arabia.
Operation Desert Storm	Operation Desert Storm (17 January 1991 – 28 February 1991) was the 2nd phase of Gulf War i.e., the combat phase.
Operation Enduring Freedom	Operation Enduring Freedom most referred to the U.S.-led combat mission in Afghanistan. The codename was also used for counter-terrorism operations in other countries targeting Al Qaeda and remnants of the Taliban.
Operation Inherent Resolve	Operation Inherent Resolve is the United States military's operational name for the international war against the Islamic State, including both a campaign in Iraq and a campaign in Syria, with a closely related campaign in Libya.
Operation Iraqi Freedom	On 20 March 2003, Operation Iraqi Freedom (OIF) began with preemptive airstrikes on former Iraqi dictator Saddam Hussein's presidential palace and selected military targets.
Operation Odyssey Dawn	Operation Odyssey Dawn was the U.S. code name for the American role in the international military operation in Libya to

	enforce United Nations Security Council Resolution 1973 during the initial period of 19–31 March 2011, which continued afterwards under NATO command as Operation Unified Protector.
Osama bin Laden	Osama bin Mohammed bin Awad bin Laden was a Saudi Arabian-born Islamic dissident and militant leader who was the founder and first general emir of Al-Qaeda from 1988 until his death in 2011.
Palmyra	Palmyra is an ancient city in south-central Syria.
Pashto	Originally spoken by the Pashtun people, Pashto became the national language of Afghanistan in 1936. It is spoken by more than 35 million people, most of whom reside in Afghanistan or Pakistan.
PSYOPS	Psychological Operations (PSYOPS) is a specialized capability and function that is integrated within various branches of the military, primarily under the United States Special Operations Command. It is a critical component of the broader information operations capabilities of the U.S. military in which activities are coordinated and conducted in accordance with strategic goals and objectives, and they play a significant role in modern warfare and national security operations.
Rabat Organization	A "Rabat organization" typically refers to the Islamic Educational, Scientific, and Cultural Organization (ISESCO), which is based in Rabat, Morocco.
Riyadh	Riyadh is the capital and largest city of Saudi Arabia.

Saddam Hussein	Saddam Hussein was an Iraqi politician and revolutionary who served as the fifth president of Iraq from 1979 to 2003.
Shatt Al-Arab	The Shatt Al-Arab is a river of some 200 kilometers (120 mi) in length that is formed at the confluence of the Euphrates and Tigris rivers in the town of Al-Qurnah in the Basra Governorate of southern Iraq.
Sima Samar	Sima Samar is a Hazara woman and human rights advocate, activist, and medical doctor within national and international forums, who served as Minister of Women's Affairs of Afghanistan from December 2001 to 2003.
Soviet–Afghan War	The Soviet–Afghan War was a protracted armed conflict fought in the Soviet-controlled Democratic Republic of Afghanistan (DRA) from 1979 to 1989.
Syed Hussain Anwari	Hussein Anwari was a Qizilbashi Tajik who was a Shi'ite and fought in the Soviet-Afghan War in the 1980s.
Syrian Democratic Forces	The Syrian Democratic Forces (SDF)[a] is a Kurdish-led coalition formed by ethnic militias and rebel groups and serves as the official military wing of the Autonomous Administration of North and East Syria (AANES).
Tal Afar	Tal Afar is a city in the Nineveh Governorate of northwestern Iraq.
Taliban	Taliban, ultraconservative political and religious faction that emerged in Afghanistan in the mid-1990s following the withdrawal of Soviet troops, the collapse of Afghanistan's communist regime, and the subsequent breakdown in civil order.

US Awards program for Justice	Rewards for Justice Program (RFJ) is a U.S. government interagency rewards program that offers money as an incentive for information leading to the arrest of leaders of terrorist groups, financiers of terrorism, including any individual that abide in plotting attacks by cooperating with foreign terrorist organizations.
Vietnam mission	A "Vietnam mission" refers to a military operation conducted during the Vietnam War (1955-1975), involving various countries aiming to counter communist forces in Vietnam. These missions included combat operations, bombing campaigns, and training efforts to support South Vietnamese forces.
War on Terrorism	War on terrorism, term used to describe the American-led global counterterrorism campaign launched in response to the terrorist attacks of September 11, 2001.

CATALOG

Leaflet 1

Title: Warning! This is just the beginning!
Imagery: A line drawing of a bomb centered on the leaflet, surrounded by Arabic text.
Objective: The leaflet aims to instill fear by warning that the bombing is only the start, linking the destruction directly to Saddam's leadership and urging the Iraqi forces to reconsider their allegiance.

Leaflet 2

Title: This site will be bombed soon.
Imagery: Another line drawing of a bomb, with text around it emphasizing the threat.
Objective: The leaflet warns of an imminent bombing of the site, urging the Iraqi soldiers to abandon their equipment and flee to save themselves, thereby weakening the enemy's defenses.

Leaflet 3

Title: It is too late!
Imagery: A line drawing showing coalition aircraft bombing Iraqi military targets, with soldiers running in panic.
Objective: This leaflet is designed to convey the inevitability of defeat, telling Iraqi soldiers that it's too late to resist, thereby demoralizing them and encouraging surrender.

Leaflet 4

Title: This site is being bombed!
Imagery: A stark line drawing of an F-117 stealth fighter dropping bombs on an Iraqi airfield, with soldiers fleeing the scene.
Objective: The leaflet aims to create immediate panic and disorder by notifying Iraqi forces that their location is under attack, compelling them to abandon their posts.

Leaflet 5

Title: Joint Forces
Imagery: This leaflet features a colorful display of various national flags surrounding the emblem of the Joint Forces.
Objective: The leaflet symbolizes unity among coalition forces, aimed at showing the strength and solidarity of the coalition against Iraq. The display of multiple flags emphasizes the global consensus against Iraq's actions, intending to demoralize Iraqi forces by highlighting their isolation on the world stage.

Leaflet 6

Title: Saddam is the only reason to bomb Iraq
Imagery: A bloodied Iraqi flag is juxtaposed with a cartoonish depiction of Saddam Hussein, reinforcing the message that he is the sole cause of the nation's suffering.
Objective: The leaflet attempts to instill a sense of betrayal among Iraqi soldiers, urging them to recognize Saddam Hussein as the reason for the bombing and devastation in Iraq. The visual portrayal of blood on the flag underlines the human cost of his leadership.

Leaflet 7

Title: Saddam's Actions Have Forced the World to War
Imagery: A black-and-white text-heavy leaflet that bluntly explains how Saddam Hussein's actions have brought Iraq into conflict with the world.
Objective: The objective is to inform Iraqi citizens and soldiers that Saddam's decisions have led to the current war, shifting the blame entirely onto him. The leaflet also emphasizes the efforts of the Joint Forces to minimize civilian casualties, aiming to create a rift between Saddam and the Iraqi people.

Leaflet 8

Title: Explosions from Above
Imagery: A stark black-and-white line drawing depicting a soldier recoiling from an intense aerial explosion, suggesting the overwhelming power of airstrikes.
Objective: The leaflet is designed to convey the devastating impact of coalition airstrikes, aiming to instill fear and encourage surrender by illustrating the futility of resisting such power.

Leaflet 9

Title: Saddam's First Line of Defense
Imagery: A simplistic line drawing showing Iraqi civilians being used as human shields in front of military installations, with tanks and soldiers positioned behind them.
Objective: The leaflet aims to expose Saddam Hussein's use of civilians as a line of defense, a strategy meant to evoke anger and resentment among the Iraqi populace and soldiers. By highlighting this tactic, the leaflet seeks to undermine the legitimacy of Saddam's rule and further demoralize his forces.

Leaflet 10

Title: The Strength of Joint Forces
Imagery: A line drawing of various aircraft, including bombers and fighters, flying in formation above a battlefield, ready to strike.
Objective: This leaflet is intended to showcase the overwhelming air superiority of the coalition forces. It serves as a stark reminder to Iraqi soldiers of the futility of resistance against such might, encouraging them to reconsider their position and surrender.

Leaflet 11

Title: You Are Isolated, Stop Resisting
Imagery: A line drawing depicting a lone Iraqi soldier surrounded by an overwhelming number of tanks, with the message urging him to stop resisting.
Objective: The leaflet's goal is to demoralize the Iraqi soldier by emphasizing their isolation and the hopelessness of their situation, encouraging them to surrender to save their lives.

Leaflet 12

Title: Surrender and Save Your Life
Imagery: The leaflet features a sequence showing an Iraqi soldier initially in combat, then shaking hands with coalition forces after surrendering.
Objective: The imagery and text work together to convey the message that surrendering is the only way to ensure survival. By showing a positive outcome for those who surrender, the leaflet aims to encourage more soldiers to lay down their arms.

Leaflet 13

Title: Oh, My Child, When Will You Return?
Imagery: A sorrowful depiction of an elderly couple, with the father thinking of their child who has been killed in battle.
Objective: This leaflet appeals to the familial bonds and emotional ties of Iraqi soldiers, reminding them of the pain and suffering their families endure. The imagery is meant to evoke guilt and encourage soldiers to desert or surrender to avoid causing further grief to their loved ones.

Leaflet 14

Title: Return Either Killed or Disabled
Imagery: A colored depiction of an Iraqi soldier envisioning his fate on the battlefield and the potential consequences of returning home either dead or disabled.
Objective: The leaflet aims to confront Iraqi soldiers with the harsh realities of war, stressing that continuing to fight will likely result in death or severe injury. The emotional appeal is designed to break the resolve of soldiers and prompt them to surrender.

Leaflet 15

Title: Saddam Leading Soldiers to Death
Imagery: A cartoonish line drawing depicting Saddam Hussein leading his soldiers towards a barrage of bombs dropped from airplanes. The soldiers are shown in disarray, panicking, and being overwhelmed by the attack.
Objective: This leaflet aims to demoralize Iraqi soldiers by portraying Saddam Hussein as leading them into certain death. The visual suggests that following Saddam results in destruction and chaos, encouraging soldiers to abandon their posts to save their lives.

Leaflet 16

Title: Surrender Now
Imagery: A line drawing of Iraqi soldiers surrendering with their hands up as they emerge from a building while tanks and jets surround them.
Objective: This leaflet aims to encourage Iraqi soldiers to surrender by showing that resistance is futile when faced with overwhelming joint forces.

Leaflet 17

Title: Inevitable Defeat
Imagery: A line drawing of an Iraqi soldier surrendering next to a tank, waving a white flag, while more soldiers in the background also surrender.
Objective: The leaflet seeks to demonstrate the inevitability of defeat and encourage surrender, emphasizing that survival is possible if the Iraqi soldiers give up.

Leaflet 18

Title: You Are Defenseless
Imagery: A textual leaflet stating, "You are defenseless!!" followed by a message about cut-off supply lines and urging Iraqi soldiers to join their Arab brothers in peace.
Objective: The leaflet is designed to demoralize Iraqi soldiers by highlighting their isolation and lack of resources, urging them to surrender for their safety.

Leaflet 19

Title: Overwhelming Force
Imagery: A line drawing of Iraqi soldiers being swept away by a powerful force, possibly representing joint forces, symbolized by a large arrow or sweep.
Objective: The leaflet seeks to illustrate the overwhelming power of the joint forces, suggesting that resistance is futile and that surrender is the only option.

Leaflet 20

Title: Relentless Bombardment
Imagery: A line drawing of Saddam Hussein surrounded by a barrage of missiles, illustrating the overwhelming firepower targeting his regime.
Objective: The leaflet aims to instill fear and hopelessness by showing the immense military pressure on Saddam's forces, encouraging surrender before it's too late.

Leaflet 21

Title: Staying here means death.
Imagery: The leaflet shows a soldier lying dead in a pool of blood while another soldier, thinking of his fate, contemplates the situation.
Objective: The message is clear and direct: remaining in the fight will lead to death. This leaflet is intended to demoralize Iraqi soldiers by reminding them of the inevitable fatal consequences of continuing to resist.

Leaflet 22

Title: Symbol of Death
Imagery: The leaflet depicts two emaciated, skeletal figures lying on the ground, with skulls in each corner of the frame. The background features a landscape with two mosque minarets. The figures appear lifeless and resigned to their fate.
Objective: This leaflet is designed to symbolize the death and destruction that awaits those who continue to fight. The use of skulls and skeletal imagery reinforces the message that persistence in battle leads only to death. The desolate figures lying on the ground serve as a visual warning to soldiers about the grim fate that awaits them if they continue to resist.

Leaflet 23

Title: Flee to Your Homes
Imagery: The leaflet shows Iraqi soldiers walking away from the battlefield, heading toward a distant, safer location. The background includes imagery of a mosque and a cityscape, representing home.
Objective: This leaflet encourages soldiers to desert and return home, emphasizing that leaving the battlefield is the safest option. The imagery of soldiers moving toward a place of safety aims to persuade them that surrendering and fleeing are in their best interest.

Leaflet 24

Title: 7th Corps Approaching
Imagery: A depiction of coalition forces, symbolized by flags of various countries, advancing toward the viewer, with an Iraqi soldier fleeing in the foreground.
Objective: The leaflet is meant to instill fear and a sense of inevitability among Iraqi forces, emphasizing that the coalition forces are closing in and that resistance is futile. It aims to encourage surrender by showing that

their defeat is imminent. It was part of the deception. It shows the 7th Corps as if they were in front of Saddam while in fact, they were moving to the flank.

Leaflet 25

Title: The Cold Face of Death
Imagery: A stark image of Saddam Hussein's face, juxtaposed with text describing death and war. The simplicity of the design focuses the viewer's attention on Saddam's role in the conflict.
Objective: This leaflet aims to associate Saddam Hussein with death and destruction, portraying him as the personification of war's grim realities. The objective is to demoralize Iraqi soldiers by making them question their loyalty to a leader who leads them to such a fate.

Leaflet 26

Title: O Sons of Iraq
Imagery: A solemn message framed with ornamental designs, urging Iraqi soldiers to consider the impact of their deaths on their homeland.
Objective: The leaflet appeals to the soldiers' sense of patriotism and familial duty, encouraging them to rethink their role in the conflict. It suggests that by dying for Saddam, they are harming Iraq's future, aiming to create internal conflict and encourage desertion.

Leaflet 27

Title: Saddam's Betrayal
Imagery: The leaflet shows Saddam Hussein straddling two soldiers as he crosses the river, with a black raven flying above, a symbol of death.
Objective: The leaflet highlights Saddam's betrayal of his soldiers, implying that he uses them to further his own ambitions while they suffer the consequences. It aims to foster resentment and disillusionment among the troops.

Leaflet 28

Title: Iranian Air Force Growing?
Imagery: The leaflet shows fighter jets marked with Iraqi and Iranian flags, hinting at growing cooperation or strength.
Objective: This leaflet is intended to create doubt and suspicion among Iraqi soldiers regarding the strength

and alliances of their military. The insinuation of an increasingly powerful Iranian air force serves to undermine confidence in their leadership and military strategy. It was meant to make Iraq wonder if their old enemy Iran would ever return those aircraft sent there for safety.

Leaflet 29

Title: Saddam's Betrayal of Shatt al-Arab
Imagery: A detailed text explanation of Saddam's decision to return the Shatt al-Arab waterway to Iran, urging Iraqi pilots to defect.
Objective: This leaflet aims to incite anger and a sense of betrayal among Iraqi soldiers by highlighting Saddam's concessions to Iran. It encourages pilots to defect, promising humane treatment and the opportunity to rebuild Iraq after the conflict.

Leaflet 30

Title: Steps to Surrender Safely
Imagery: The leaflet provides a step-by-step guide with illustrations on how to safely surrender to coalition forces, showing a soldier following these steps.
Objective: This leaflet is a practical guide for Iraqi soldiers on how to surrender safely. It aims to reduce resistance by making the process of surrendering clear and straightforward, reassuring soldiers that they will be treated humanely if they follow the instructions.

Leaflet 31

Title: Leaflet Depicting Tuned Radio
Imagery: The leaflet shows soldiers tuning a radio, with text instructing them to tune to a specific frequency.
Objective: This leaflet aims to inform Iraqi soldiers of the designated radio frequency for coalition broadcasts. By ensuring they are tuned in, the coalition aims to maintain communication and control over the information the soldiers receive.

Leaflet 32

Title: Radio Kuwait Broadcast
Imagery: The leaflet features a simple drawing of a radio antenna with a list of FM frequencies alongside the emblem of Kuwait.
Objective: The leaflet is intended to direct the Iraqi population and soldiers to listen to Radio Kuwait, which

likely broadcasts coalition propaganda and information, aiming to sway public opinion and lower morale among Iraqi forces.

Leaflet 33

Title: O Citizens of Basra!
Imagery: The leaflet features a straightforward text warning the citizens of Basra about the imminent destruction of their city and urging them to leave immediately.
Objective: This leaflet is intended to create panic and prompt evacuation, thereby minimizing civilian casualties while also disrupting Iraqi military operations by causing confusion and chaos among the population.

Leaflet 34

Title: Second 'O Citizens of Basra!' Warning
Imagery: The leaflet contains a detailed text explaining the consequences of Saddam Hussein's actions and urging the citizens of Basra to flee the city.
Objective: Similar to the previous leaflet, this one aims to incite fear and cause the evacuation of Basra, emphasizing the inevitability of coalition military action and the danger of remaining in the city.

Leaflet 35

Title: Stay in Your Homes!!
Imagery: The leaflet is text-heavy, written in English, instructing civilians to stay indoors and avoid coalition forces to stay safe.
Objective: The purpose of this leaflet is to ensure civilian safety by urging them to stay away from the conflict zones. It also seeks to prevent accidental harm to civilians, which could damage the coalition's image.

Leaflet 36

Title: Joint Forces Warning to Iraqi Citizens
Imagery: The leaflet is text-based and advises Iraqi citizens to remain in their homes and avoid approaching joint forces to stay safe during impending military actions.
Objective: This leaflet aims to minimize civilian casualties during military operations by clearly instructing them to stay indoors. It serves as a precautionary measure to prevent any unintentional harm to non-combatants.

Leaflet 37

Title: Anti-Saddam Messages on Signs
Imagery: The leaflet shows a map of Iraq with signs carrying messages against Saddam and promoting peace, with a dove holding an olive branch.
Objective: The leaflet aims to portray Saddam as a warmonger and to encourage Iraqi citizens to support peace by rejecting his leadership. The dove and olive branch symbolize the coalition's desire for peace and the end of Saddam's reign.

Leaflet 38

Title: Dear Iraqi Citizens – Reject Saddam
Imagery: The leaflet features text urging Iraqi citizens to reject Saddam Hussein's leadership and join the coalition for peace.
Objective: The objective is to erode support for Saddam among Iraqi citizens, encouraging them to see the coalition as a liberating force rather than an occupying one. The leaflet seeks to foster dissent within Iraq against Saddam's regime.

Leaflet 39

Title: My Iraqi Brother, What We Want is Peace
Imagery: The leaflet features a simple handwritten message stating that the desire is for peace.
Objective: The leaflet is designed to appeal to the shared humanity of Iraqi soldiers and citizens, emphasizing that the coalition's primary goal is peace, not conflict. This is meant to reduce resistance and encourage cooperation with coalition forces.

Leaflet 40

Title: Flags of Iraq and Saudi Arabia
Imagery: The leaflet depicts the flags of Iraq and Saudi Arabia with two figures shaking hands, set against the backdrop of a rising sun.
Objective: This leaflet aims to symbolize a potential alliance or peaceful relations between Iraq and Saudi Arabia. It suggests a future where cooperation replaces conflict, encouraging Iraqi forces and citizens to consider a peaceful resolution.

Leaflet 41

Title: Peace or Conflict?
Imagery: A colored illustration showing two soldiers shaking hands, with thought bubbles above their heads. One bubble shows a smiling face of a Middle Eastern leader, while the other bubble shows a face that appears stern and unfriendly. The background is a desert landscape with the sun shining above.
Objective: The leaflet is designed to promote unity and reconciliation between different factions or armies, encouraging soldiers to choose peace over conflict. The contrasting expressions in the thought bubbles suggest a choice between two paths: one leading to cooperation and mutual understanding, and the other leading to division and hostility. The handshake signifies a peaceful resolution, urging soldiers to lay down their arms and cooperate rather than continue fighting.

Leaflet 42

Title: In Peace, We Will Always Remain United
Imagery: The leaflet features simple Arabic text on a plain background that reads, "In peace, we will always remain united."
Objective: The leaflet is meant to reinforce the idea of unity and solidarity among people, emphasizing that peace is the key to maintaining strong and harmonious relationships. By focusing solely on the message without any complex imagery, the leaflet seeks to convey a straightforward and powerful statement encouraging people to choose peace to preserve unity.

Leaflet 43

Title: Warning!
Imagery: The leaflet depicts three bombs falling from the sky towards a city with minarets. Above the bombs is a skull and crossbones symbol, and the word "Warning" is prominently displayed in red.
Objective: This leaflet serves as a clear and urgent warning to avoid suspicious objects that might be explosive. The imagery of bombs and the skull and crossbones symbol are designed to instill fear and caution, urging civilians and soldiers to report any strange objects immediately to prevent harm.

Leaflet 44

Title: Invitation Card
Imagery: The leaflet features an emblem of the Joint Forces Command with text inviting Iraqi soldiers to join the coalition forces. The emblem is encircled by olive branches and is accompanied by a red ribbon.

Objective: This leaflet serves as an "invitation card" to Iraqi soldiers, offering them safety, security, and medical care if they choose to join the coalition forces. The leaflet attempts to undermine Saddam Hussein's authority by presenting the coalition as a more favorable option, thereby encouraging defection and weakening enemy forces.

Leaflet 45

Title: An Offer of Safe Haven
Imagery: A scene depicting soldiers relaxing and sharing a meal in a camp, with a Saudi Arabian flag in the background. In the second frame, an Iraqi soldier is shown receiving a pass from a coalition soldier.
Objective: The leaflet aims to entice Iraqi soldiers to surrender peacefully by showing them the benefits of defecting to the coalition forces. The imagery of soldiers being treated well in the camp, along with the friendly exchange between soldiers, highlights the advantages of joining the coalition, promising humane treatment and safety.

Leaflet 46

Title: Geneva Convention Protections
Imagery: The leaflet shows text outlining the protections offered by the Geneva Convention, including humane treatment, food and water, medical care, shelter, and a return home after the war.
Objective: The leaflet reassures Iraqi soldiers that they will be treated according to international laws if they surrender. By emphasizing the humane treatment and basic needs they will receive, the leaflet encourages soldiers to lay down their arms and seek protection under the Geneva Convention.

Leaflet 47

Title: Choose Peace or Continue Fighting?
Imagery: The leaflet shows two soldiers shaking hands in a desert, with thought bubbles depicting Saddam Hussein's face in one and a hopeful expression in the other. The scene is split, showing both cooperation and hesitation.
Objective: The leaflet is designed to provoke a decision between continuing to fight under Saddam's regime or choosing peace and cooperation. The visual contrast between the two thought bubbles serves to highlight the benefits of peace over the uncertainty and fear associated with continued conflict.

Leaflet 48

Title: Desert Storm (Sahara Whirlwinds) is coming to your area. Flee immediately.
Imagery: A line drawing depicting an overwhelming storm of aircraft and military vehicles advancing, representing the unstoppable force of the coalition.
Objective: The leaflet aims to instill immediate fear and urgency, encouraging Iraqi soldiers and civilians to flee the area to avoid the impending military onslaught.

Leaflet 49

Title: Staying here means death.
Imagery: The leaflet shows a soldier lying dead in a pool of blood while another soldier imagines the grim fate awaiting those who stay.
Objective: The leaflet warns Iraqi soldiers that remaining in their current position is suicidal, encouraging them to desert their posts to save their lives.

Leaflet 50

Title: CEASE RESISTANCE--BE SAFE
Imagery: Text-heavy leaflet providing instructions for surrendering safely, including removing the magazine from the weapon, slinging the weapon over the left shoulder, and approaching coalition forces with hands raised.
Objective: This leaflet provides a clear and detailed process for Iraqi soldiers to surrender without being harmed, promoting safe defection and reducing resistance.

Leaflet 51

Title: Iraqi citizens, Saddam's military has placed your lives in danger.
Imagery: A text-heavy leaflet with red borders, urging civilians to evacuate and move northward to avoid being caught in the crossfire as coalition forces advance.
Objective: The leaflet seeks to minimize civilian casualties by warning them to leave the area immediately, emphasizing the coalition's intention to avoid harming innocent lives.

Leaflet 52

Title: Imagery and Objective: A visual representation of surrender and family reunification.
Imagery: A drawing split into two parts, with one showing a soldier surrendering peacefully, and the other depicting the soldier reuniting with his family, under the flags of coalition nations.
Objective: This leaflet is designed to encourage Iraqi soldiers to surrender by appealing to their desire to return home safely and reunite with their families.

Leaflet 53

Title: PRISONER OF WAR HOLDING COMPOUND RULES AND REGULATIONS
Imagery: A straightforward list of rules for prisoners of war, emphasizing obedience, cleanliness, and the avoidance of contraband and prohibited activities.
Objective: The leaflet is intended to inform Iraqi prisoners of war about the rules they must follow to ensure their safety and maintain order within the holding compound.

Leaflet 54

Title: My brother Iraqi soldiers, your resistance distracted the enemies of Iraq.
Imagery: A drawing of Iraq with fists rising in defense, symbolizing strength and unity against invaders.
Objective: The leaflet serves to boost the morale of Iraqi soldiers, emphasizing their importance in the resistance and encouraging continued loyalty to the Iraqi regime.

Leaflet 55

Title: My brother, the Arab soldier.
Imagery: A drawing of Arab soldiers on the battlefield with a thought bubble showing a common heritage, questioning their loyalty to the cause.
Objective: The leaflet aims to create doubt and guilt among Arab soldiers fighting against Iraq, suggesting that they are betraying their own heritage and people.

Leaflet 56

Title: Allah Akbar. God is Great.
Imagery: A symbolic image of a handshake between forces labeled as "Arabic and Islamic Forces," with a fallen American soldier in the background.

Objective: The leaflet's goal is to unite Arab and Islamic forces under a common cause against a foreign invader, strengthening their resolve and solidarity.

Leaflet 57

Title: PRISONER OF WAR HOLDING COMPOUND RULES AND REGULATIONS
Imagery: The leaflet contains detailed text outlining the rules and regulations for prisoners of war, emphasizing discipline and order within the compound.
Objective: The leaflet seeks to establish clear expectations and maintain order among prisoners of war, ensuring compliance and reducing the likelihood of unrest.

Leaflet 58

Title: Time isn't on your side... How many days are left?... Keep counting... until January 15th!!
Imagery: A cartoonish image of an Iraqi soldier and a civilian looking at a calendar with panic, counting down the days.
Objective: The leaflet uses a countdown to create urgency and panic, suggesting that the time is running out for the Iraqi forces and encouraging them to surrender before the deadline.

Leaflet 59

Title: Oops
Imagery: A cartoon drawn for an American Newspaper showing Saddam Hussein comically attempting to behead a figure representing Iraq, but instead, his efforts backfire, with the sword missing its mark and the blade accidentally harming himself, signified by an "Oops" thought bubble.
Objective: The leaflet mocks Saddam Hussein's leadership, portraying his actions as self-destructive and incompetent. The cartoon aims to demoralize his supporters by highlighting the futility and recklessness of his decisions. It was printed by accident and was not meant for distribution.

Leaflet 60

Title: 30 February 1991
Imagery: The leaflet shows the date "30 February 1991" with accompanying text that refers to Saddam Hussein's prediction.
Objective: The leaflet mocks Saddam Hussein by implying that his promises are as impossible as the date "30

February." The intention is to undermine the credibility of Saddam's assurances to solve the problem of Kuwait, suggesting that they are empty promises.

Leaflet 61

Title: Activities Scheduled
Imagery: The leaflet lists a detailed daily schedule of activities, including waking up, meals, medical clinic visits, and attendance counting, along with rules for prisoners of war.
Objective: The leaflet aims to inform prisoners of war about the regimented daily schedule they are expected to follow, ensuring order and discipline within the camp. The clear schedule also serves to emphasize the structured and controlled environment in which the prisoners are kept, reinforcing the authority of the detaining forces.

Leaflet 62

Title: Yesterday, we presented to you proof...
Imagery: The leaflet features a photograph of a military aircraft dropping bombs, symbolizing the overwhelming power and capability of the Joint Forces. The text reinforces this message, emphasizing that this is a second opportunity for the Iraqi forces to surrender and save their lives.
Objective: The leaflet aims to psychologically pressure the Iraqi soldiers by demonstrating the unstoppable power of the Joint Forces and offering them a chance to surrender before it's too late. This is intended to weaken enemy morale and encourage desertion.

Leaflet 63

Title: Bombing Warning
Imagery: A B-52 bomber—in service since the Vietnam War—dropping bombs with text warning that the 48th Infantry Division will be bombed again tomorrow.
Objective: This leaflet serves as a follow-up warning to a previous leaflet, stating that the promised bombing of the 48th Infantry Division has already taken place and will be repeated. The objective is to instill fear and encourage the soldiers to desert their positions before the next bombing.

Leaflet 64

Title: 48th Infantry Division Bombing
Imagery: A simple text message framed by a black border, warning that the 48th Infantry Division will be

bombed again. The leaflet emphasizes the fulfillment of a previous warning and the inevitability of further bombing.
Objective: This leaflet is a psychological tool meant to increase anxiety and fear among the soldiers of the 48th Infantry Division. It reinforces the reality of the bombing that already occurred, promising that it will happen again. The aim is to weaken their resolve and push them towards surrender or desertion, presenting the choice as one between death and survival by joining the Joint Forces.

Leaflet 65

Title: 20th Infantry Division Bombing
Imagery: A similar style to the previous leaflets, with a red-bordered warning addressed to the 20th Infantry Division about their impending bombing.
Objective: This leaflet targets the 20th Infantry Division, informing them that their position was bombed as promised, and it will happen again. It aims to weaken their morale and force them to surrender or desert their positions to avoid death.

Leaflet 66

Title: For Your Safety
Imagery: Text providing instructions for Iraqi naval forces on how to safely surrender to coalition forces, including raising a flag and avoiding threatening movements.
Objective: This leaflet instructs Iraqi naval personnel on how to surrender safely to coalition forces. It emphasizes the importance of following specific procedures to ensure their safety, aiming to reduce resistance and encourage surrender.

Leaflet 67

Title: Last Line of Defense
Imagery: Text in Arabic surrounded by a decorative border, warning Iraqi soldiers that they are the last line of defense between Saddam and the coalition forces.
Objective: The leaflet seeks to demoralize Iraqi soldiers by emphasizing the overwhelming force they face and urging them not to share Saddam's fate. It encourages them to surrender rather than resist, suggesting that their loyalty to Saddam is misplaced and will lead to their destruction.

Leaflet 68

Title: Overwhelmed by Leaflets
Imagery: A cartoon drawing of Saddam Hussein angrily shaking his fist as he is buried under a massive pile of leaflets. An aircraft is seen dropping more leaflets from above, symbolizing the overwhelming amount of propaganda being disseminated.
Objective: This leaflet is intended to mock and demoralize Saddam Hussein by depicting him as being literally buried under the propaganda efforts of the coalition forces. The imagery suggests that no matter how much he resists, he cannot escape the overwhelming influence and message of the coalition, which aims to erode his authority and credibility among his troops and the Iraqi population. This leaflet was not used for distribution but designed for boosting morale of American PSYOP soldiers.

Leaflet 69

Title: Massive Bomb about to Drop
Imagery: This leaflet illustrates a giant bomb being dropped from a plane, emphasizing the massive and unavoidable destruction awaiting those who do not surrender.
Objective: The objective is to terrorize the Iraqi soldiers into surrendering by depicting the sheer scale of the coalition's bombing capability. It suggests that resistance is futile and that surrender is the only way to avoid complete annihilation.

Leaflet 70

Title: No! Not for you!
Imagery: The leaflet depicts an Iraqi civilian rejecting an offering from soldiers, with children eating in the background. The scene questions the authenticity of Arab customs and traditions under Saddam's regime.
Objective: This leaflet aims to create distrust between Iraqi soldiers and the regime by highlighting the disparity between the treatment of soldiers and civilians. It questions the legitimacy of the regime's actions, suggesting that they betray Arab customs and traditions. The goal is to weaken the soldiers' resolve and loyalty to Saddam.

Leaflet 71

Title: We can live for 20 years
Imagery: The leaflet depicts a military figure standing confidently, surrounded by poorly dressed and malnourished children, emphasizing the disparity between the promises of the regime and the actual

conditions.

Objective: This leaflet aims to highlight the harsh realities faced by Iraqi citizens under Saddam's rule, contrasting the regime's empty promises with the visible suffering of the people. It seeks to undermine the morale of Iraqi soldiers by showing the stark difference between what they are told and the actual living conditions.

Leaflet 72

Title: Can you take care of your family when you are physically disabled?

Imagery: An old image used as early as WWII shows an injured soldier, standing on crutches with a bandaged eye and missing a leg. The image conveys the physical toll of the war.

Objective: This leaflet is designed to discourage Iraqi soldiers from continuing to fight by making them question whether they can support their families if they are injured or disabled in the war. It plays on the fear of permanent disability and the inability to fulfill their family obligations.

Leaflet 73

Title: Saddam frying soldiers

Imagery: The leaflet illustrates a caricature of Saddam Hussein cooking soldiers in a frying pan over a large fire, with one soldier about to be placed into the pan. The image is grotesque and exaggerated, typical of propaganda art.

Objective: This leaflet aims to depict Saddam as a tyrant who is sacrificing his own soldiers for his personal gain. The imagery is meant to instill fear and demoralize the Iraqi soldiers, making them feel that they are expendable and that their lives are being wasted by a cruel dictator.

Leaflet 74

Title: Tanks and planes in chaos

Imagery: The leaflet depicts a chaotic battlefield where tanks are being destroyed by airstrikes, with planes swooping down and firing missiles. The image is dynamic and filled with action, conveying the intensity of the conflict.

Objective: This leaflet seeks to highlight the overwhelming firepower and technological superiority of the coalition forces. It serves as a warning to Iraqi soldiers about the futility of resisting and the inevitability of defeat if they continue to fight.

Leaflet 75

Title: Flood of soldiers
Imagery: The leaflet shows a powerful wave of coalition soldiers, tanks, and military vehicles overwhelming and drowning the Iraqi forces. The imagery is symbolic of an unstoppable force crashing down on those who try to resist.
Objective: This leaflet is designed to emphasize the sheer strength and numbers of the coalition forces, conveying the message that resistance is futile. It aims to demoralize Iraqi soldiers by making them feel powerless against the overwhelming might of the coalition.

Leaflet 76

Title: Crumbling foundation of power
Imagery: The leaflet illustrates Saddam standing atop a large, crumbling structure labeled "الأكاذيب" (Lies), with cracks forming as it begins to collapse. Beneath the structure, a crowd of soldiers appears to be holding it up, struggling to keep it from falling apart.
Objective: This leaflet seeks to expose the foundation of Saddam's power as being built on lies and deception. It conveys the message that his regime is on the verge of collapse and that continuing to support it is both futile and dangerous. The leaflet aims to persuade soldiers to abandon Saddam before they are crushed along with his regime.

Leaflet 77

Title: Worthless currency
Imagery: The leaflet depicts a shopkeeper handing over a roll of Saddam's currency to a customer, with the caption "على الأقل لها قيمة الآن" (At least it has value now). The currency, represented as toilet paper, symbolizes its worthlessness.
Objective: This leaflet aims to highlight the severe devaluation of Iraq's currency under Saddam's regime, suggesting that it is now as worthless as toilet paper. The message is intended to undermine confidence in Saddam's leadership by illustrating the economic collapse and the dire situation in Iraq.

Leaflet 78

Title: Green Creep
Imagery: The leaflet shows a sinister-looking figure identified as the "Green Creep" (زحف الأخضر) burning a document, with the words "جلاد بغداد" (The Executioner of Baghdad) prominently displayed. The figure's

menacing stance suggests danger and fear.
Objective: This leaflet portrays the "Green Creep" as a destructive force associated with Baghdad's executioner, likely alluding to the oppressive and violent actions of Saddam's regime. The imagery and text work together to evoke fear and encourage defection or resistance against the regime.

Leaflet 79

Title: The meal that Saddam gives you
Imagery: The leaflet displays a large question mark followed by a list of meals provided by the coalition forces. The Arabic text at the top translates to "The meal that Saddam gives you?"
Objective: This leaflet contrasts the scarcity of food under Saddam's regime with the nutritious meals provided by the coalition forces, aiming to highlight the dire living conditions in Iraq and to persuade Iraqi soldiers or civilians of the benefits of cooperating with the coalition.

Leaflet 80

Title: The value of your money
Imagery: The leaflet compares the Iraqi currency to a worthless piece of paper, symbolizing the severe economic decline. The text discusses the lack of food, medicine, and hygiene supplies in Iraq, painting a bleak picture of daily life under Saddam's rule.
Objective: The leaflet seeks to undermine confidence in Saddam's government by emphasizing the economic hardships faced by ordinary Iraqis. It encourages the people to rise against Saddam, highlighting the futility of supporting a regime that has led to widespread famine and poverty.

Leaflet 81

Title: Economic Collapse and Despair in Iraq
Imagery: The leaflet contains a rectangular block of Arabic text, bordered by a black line. The text is presented in a formal, serif font.
Objective: The leaflet emphasizes the severe economic and humanitarian crisis in Iraq, highlighting the loss of monetary value, shortages of food, and widespread misery. It squarely places the blame on Saddam Hussein's regime and calls on the people to act against him to prevent further suffering.

Leaflet 82

Title: Imminent Bombing of the 16th Infantry Division
Imagery: The leaflet features an image of a military aircraft dropping bombs, accompanied by a block of Arabic text. The background is slightly darkened, possibly to simulate a night operation.
Objective: This leaflet warns the 16th Infantry Division of an imminent bombing scheduled for the next day. It urges the soldiers to abandon their positions immediately to save their lives. The use of imagery reinforces the seriousness of the threat, aiming to instill fear and prompt desertion.

Leaflet 83

Title: The Promise Fulfilled: Repeat Bombing Threat
Imagery: The leaflet is similar in style to #82 but uses a red background for its text block. The same military aircraft is shown, but the text is emphasized with the use of the red color, likely to underline the urgency.
Objective: The leaflet reiterates the warning from leaflet #82, emphasizing that the promised bombing has already occurred, and the threat will be repeated. It offers soldiers a stark choice: face death or accept the Joint Forces' invitation to surrender and survive.

Leaflet 84

Title: The Burden of Saddam: Eleven Years of Suffering
Imagery: The leaflet depicts a cartoonish scene in full color. A soldier, burdened by a much larger figure representing Saddam Hussein, is seen struggling under the weight. The larger figure appears to be exerting dominance over the smaller, exhausted soldier.
Objective: This leaflet aims to demoralize Iraqi soldiers by highlighting the prolonged burden and suffering caused by Saddam Hussein's leadership. The message suggests that continuing to support Saddam is futile and that the soldiers are being exploited.

Leaflet 85

Title: The Truth: The Futility of War and a World Against Saddam
Imagery: This leaflet contains a block of Arabic text against a plain background, formatted similarly to official documents. The text is centered, with bold headers and regular font for the body text.
Objective: The leaflet reflects on the eight-year war against Iran, calling it a fruitless endeavor that cost half a million lives. It warns that a new generation is at risk and that the world is uniting against Saddam. The message is a call to the soldiers to not allow Saddam to continue leading them to destruction.

Leaflet 86

Title: Saddam's Aggression: A Call to Surrender
Imagery: The leaflet features another block of Arabic text, styled similarly to leaflet #85, with bold headers and regular text. The background is neutral, focusing attention solely on the message.
Objective: The leaflet directly addresses the Iraqi military, blaming Saddam's aggressive policies for the current bombings. It explains that the bombing is targeted only at military objectives and that resistance is futile. The leaflet encourages soldiers to surrender and save their lives, placing the blame for the conflict squarely on Saddam.

Leaflet 87

Title: Safe Conduct Pass: Instructions for Surrender
Imagery: This leaflet presents a block of Arabic text framed as an official-looking document. It uses a structured, ordered format to list specific instructions.
Objective: The leaflet serves as a "Safe Conduct Pass," providing instructions for surrendering safely to coalition forces. It outlines steps for Iraqi soldiers to follow to avoid death and be treated well upon surrender. The message aims to facilitate the surrender process by assuring soldiers of their safety and humane treatment.

Leaflet 88

Title: Safe Conduct Card: Protection Under the Geneva Convention
Imagery: The leaflet features a detailed, official-looking card with various national flags and a prominent emblem of the Joint Forces. The text is structured like an official document, with headings and signatures.
Objective: The leaflet functions as a "Safe Conduct Card," granting the bearer permission to cross the border and surrender to coalition forces. It reassures soldiers of their safety and promises fair treatment under the Geneva Convention. The card is designed to encourage surrender by offering security and protection.

Leaflet 89

Title: No Escape
Imagery: A burning vehicle, overturned, amidst a barren landscape, with a thick plume of black smoke rising.
Objective: The message aims to depict the inescapable destruction awaiting those who resist or try to flee, reinforcing the inevitability of defeat.

Leaflet 90

Title: Mined Escape Routes
Imagery: A desolate field with scattered landmines, highlighting the danger of hidden explosives.
Objective: This leaflet is designed to warn Al-Qaeda members that their escape routes are perilous and likely to lead to death, emphasizing the futility of attempting to evade capture or attack.

Leaflet 91

Title: No Place to Hide
Imagery: A split image featuring Osama bin Laden: one side shows him as a wanted fugitive surrounded by dollar bills, while the other shows him behind bars.
Objective: The leaflet conveys that no matter how far a criminal like Osama bin Laden runs, he will ultimately be captured and brought to justice. It emphasizes the certainty of retribution.

Leaflet 92

Title: Escape to Survive
Imagery: Two images side by side; one shows a family, while the other depicts a lone man standing beside a vehicle in a barren landscape.
Objective: The leaflet urges individuals to escape with their families to ensure their survival, contrasting the safety of family togetherness with the isolation and danger of staying behind.

Leaflet 93

Title: The Danger of Mines
Imagery: A bright red triangular warning sign with a skull and crossbones, surrounded by images of various types of mines and bombs.
Objective: The leaflet aims to warn civilians, particularly children, of the dangers of unexploded ordnance. It emphasizes the lethal risks associated with touching or playing near mines.

Leaflet 94

Title: In Our Crosshairs
Imagery: A black-and-white photo of Taliban or Al-Qaeda members with crosshairs superimposed over their faces.

Objective: This leaflet serves as a direct threat, stating that these members are the targets of military operations. It aims to intimidate and discourage further involvement in insurgency activities.

Leaflet 95

Title: Betrayed by Leaders
Imagery: A collage featuring the faces of Osama bin Laden and other key figures, overlaid with a ghostly skull and images of armed militants.
Objective: The leaflet questions why individuals would protect those who have betrayed them, suggesting that these leaders have abandoned their followers. It is designed to sow distrust and encourage defection.

Leaflet 96

Title: Terror's Face
Imagery: The leaflet shows images of Osama bin Laden, a skull, and a group of armed men, creating a fearsome and ominous atmosphere.
Objective: To convey the terror and fear that these individuals bring, emphasizing that they are the ones who send people to die in war.

Leaflet 97

Title: Bin Laden Abandons You
Imagery: A dead body lies on the ground next to a block of text denouncing Osama bin Laden as a murderer and coward.
Objective: To demoralize fighters by portraying Osama bin Laden as a coward who abandoned his followers, urging them to surrender before they suffer the same fate.

Leaflet 98

Title: Hamid Karzai's Leadership
Imagery: A portrait of Hamid Karzai alongside text identifying him as the head of the interim government of Afghanistan.
Objective: To legitimize the authority of Hamid Karzai and the interim Afghan government, encouraging support and cooperation.

Leaflet 99

Title: Terrorists Betray You
Imagery: Images of terrorists, including Osama bin Laden, alongside text that labels them as people who do not care about the lives or families of others.
Objective: To dissuade individuals from supporting or following terrorists by highlighting their disregard for human life and family.

Leaflet 100

Title: Questioning the Fight
Imagery: A group of men, including Osama bin Laden, alongside a block of text questioning why they are taking others to fight in a war with no clear purpose.
Objective: To create doubt and discourage involvement in the conflict, questioning the motives behind fighting for terrorists.

Leaflet 101

Title: Suffering Brought by Terrorists
Imagery: An image of a woman wrapped in a blanket during a harsh winter, accompanied by a message that the Taliban and Al-Qaeda bring only pain and suffering to Afghanistan.
Objective: To highlight the negative impact of the Taliban and Al-Qaeda on the Afghan people, urging them to reject these groups.

Leaflet 102

Title: Surrender to Save Yourself
Imagery: A Taliban fighter holding a weapon, with text urging surrender to the coalition forces.
Objective: To encourage Taliban and Al-Qaeda fighters to surrender to the coalition forces, suggesting that resistance is futile.

Leaflet 103

Title: Life of Fear
Imagery: A bleak image of two individuals in a dark environment, overlaid with images of Osama bin Laden and a skull, symbolizing death and fear.

Objective: To emphasize that life under terrorists is filled with fear and suffering, urging people to distance themselves from such groups.

Leaflet 104

Title: Life of Peace
Imagery: In contrast to the previous leaflet, this one shows a peaceful and prosperous scene, with families and children, overlaid with a message about the benefits of a life without terrorists.
Objective: To promote the idea that rejecting terrorism leads to peace and prosperity, offering a hopeful alternative to life under terrorist influence.

Leaflet 105

Title: Osama bin Laden Abandons You
Imagery: The leaflet features an image of a man dressed in a suit, identified as Osama bin Laden. The text accuses him of abandoning his followers and being a "honorless murderer."
Objective: The objective is to demoralize Osama bin Laden's followers by portraying him as a dishonorable leader who has deserted them. The leaflet seeks to sow doubt and fear among those who might still be loyal to bin Laden, encouraging them to surrender.

Leaflet 106

Title: Zawahiri: The Honorless Murderer
Imagery: This leaflet shows a depiction of Ayman al-Zawahiri, sitting on a pile of bodies, with other images depicting mourning and despair. The text labels Zawahiri as a "honorless murderous man."
Objective: The leaflet aims to undermine Zawahiri's credibility by depicting him as a cold-blooded murderer who causes only death and destruction. It seeks to dissuade support for Zawahiri by emphasizing his lack of honor and morality.

Leaflet 107

Title: No Escape from Coalition Bombing
Imagery: The leaflet shows an image of Osama bin Laden's face fading into the background, with a coalition aircraft dropping bombs. The accompanying text warns there is no possibility of escape or hiding.
Objective: The objective is to instill fear and hopelessness in those targeted by coalition forces by emphasizing the inevitability of being caught or destroyed, no matter where they try to hide.

Leaflet 108

Title: Taliban and Al-Qaeda Destroy Your Cities
Imagery: The leaflet displays before-and-after images of cities, showing destruction attributed to the Taliban and Al-Qaeda. The text blames these groups for the devastation.
Objective: This leaflet seeks to turn public opinion against the Taliban and Al-Qaeda by holding them responsible for the destruction of their homes and cities. It encourages people to distance themselves from these groups.

Leaflet 109

Title: Stop the Fight!
Imagery: The leaflet features a cartoonish image of a soldier handing over a weapon to another person, symbolizing the end of a fight. A distressed figure is depicted in the background.
Objective: The leaflet aims to encourage the cessation of fighting and promote peace. It appeals to individuals to stop the violence and surrender to avoid further conflict.

Leaflet 110

Title: Afghanistan: An Islamic Republic
Imagery: The leaflet depicts a scene where a civilian is engaging in dialogue with an official figure, with the Afghan flag prominently displayed. The text emphasizes that Afghanistan is an Islamic Republic, not a Mullahcracy.
Objective: The objective is to affirm the identity of Afghanistan as an Islamic Republic, countering any narrative that the country should be governed by religious extremists. It seeks to bolster national pride and support for the legitimate government.

Leaflet 111

Title: Reward for Bin Laden and Zawahiri
Imagery: The leaflet presents photos of Osama bin Laden and Ayman al-Zawahiri with a caption offering a $25 million reward for information leading to their capture.
Objective: The objective is to incentivize individuals to provide information on the whereabouts of these high-profile targets by offering a substantial monetary reward.

Leaflet 112

Title: Reward Information for Bin Laden and Zawahiri
Imagery: Similar to leaflet 111, this leaflet displays the images of bin Laden and Zawahiri, reiterating the reward of up to $25 million for information on their location.
Objective: The leaflet reinforces the message that providing information about bin Laden and Zawahiri can lead to a significant financial reward, aiming to encourage collaboration with coalition forces.

Leaflet 113

Title: Coalition Forces Block Humanitarian Aid
Imagery: The leaflet shows a militant holding a weapon while humanitarian aid is being blocked. The text warns that the coalition forces' attack prevents humanitarian aid from reaching the region.
Objective: This leaflet aims to shift the blame for the lack of humanitarian aid onto the coalition forces, likely as part of a psychological warfare tactic to turn the local population against them.

Leaflet 114

Title: Provide Information to Continue Receiving Aid
Imagery: The leaflet depicts a scene where civilians are receiving humanitarian aid, with a message encouraging the provision of information about the Taliban, Al-Qaeda, and Gulbuddin Hizb to coalition forces to continue receiving aid.
Objective: The objective is to incentivize the population to share intelligence with coalition forces by linking it to the continuation of humanitarian aid. The leaflet aims to gather actionable intelligence while maintaining support from the local population.

Leaflet 115

Title: Help the Commonwealth Nations
Imagery: The leaflet shows a coalition soldier interacting with smiling children and local families, conveying a positive image of cooperation and assistance.
Objective: The objective is to encourage local support for the Commonwealth Nations' efforts to restore peace and stability in Afghanistan. It seeks to build trust and collaboration between the coalition forces and the local population.

Leaflet 116

Title: Do Not Interfere with Coalition Forces
Imagery: This leaflet features a military convoy and helicopters, with text urging civilians not to interfere with coalition personnel and equipment.
Objective: The leaflet aims to prevent civilians from getting involved in military operations, thereby reducing the risk of harm to themselves and ensuring the smooth operation of coalition activities.

Leaflet 117

Title: Afghanistan Government and Coalition Aid
Imagery: The leaflet displays scenes of humanitarian assistance provided by Afghan government and coalition forces, showing them distributing aid to the local population.
Objective: The leaflet's objective is to reinforce the message that the Afghan government and coalition forces are working together to provide humanitarian aid, encouraging locals to accept and support their efforts.

Leaflet 118

Title: Report Attacks on Coalition Forces
Imagery: This leaflet shows images of coalition forces interacting with locals and providing aid, with text urging the reporting of any attacks on these forces.
Objective: The leaflet seeks to mobilize the local population to protect coalition forces by reporting any attacks, thereby ensuring continued humanitarian aid and support.

Leaflet 119

Title: Combat Smuggling in Iraq
Imagery: The leaflet features an image of arrested individuals involved in smuggling Iraq's natural resources, with warnings of severe punishment.
Objective: The objective is to deter smuggling activities by highlighting the severe legal consequences. It encourages locals to report illegal activities to the authorities, reinforcing the rule of law.

Leaflet 120

Title: Punishment for Smuggling
Imagery: This leaflet displays images of convicted smugglers in prison, with details of their crimes and

penalties.
Objective: The leaflet aims to serve as a warning to those involved in smuggling activities. By showcasing the punishment, it seeks to deter similar crimes and promote law enforcement.

Leaflet 121

Title: Support the New Afghan Government
Imagery: The leaflet depicts the new Afghan government as a source of freedom and progress, with imagery showing positive community interactions.
Objective: The objective is to encourage local support for the new government, positioning it as a provider of new freedoms and opportunities for the Afghan people.

Leaflet 122

Title: Stay and Die
Imagery: The leaflet features a stark contrast between scenes of violence and the potential for peace, with the text "Stay and Die."
Objective: The leaflet's message is clear: those who continue to resist will face death, while surrender and cooperation offer a chance for survival and peace.

Leaflet 123

Title: Death is Inevitable
Imagery: The leaflet shows skulls and grim imagery to emphasize the message that attacking Afghanistan will lead to certain death.
Objective: The objective is to dissuade combatants from attacking Afghanistan by portraying the inevitability of death as a consequence. It seeks to instill fear and discourage hostile actions.

Leaflet 124

Title: We Stop Your Vehicles for Security
Imagery: The leaflet depicts coalition forces stopping vehicles in an area where enemies of Afghanistan have been detected, with a focus on maintaining peace and security.
Objective: The objective is to justify the actions of coalition forces in stopping vehicles, ensuring the local population understands that these actions are for their safety. It encourages cooperation and reporting of any enemy activity.

Leaflet 125

Title: America's Assistance
Imagery: This leaflet shows Afghan children, sacks of food aid labeled with "USA," and a man carrying one of these sacks on his back.
Objective: To highlight the substantial financial assistance provided by America to Afghanistan, stressing the U.S.'s role in helping the Afghan people.

Leaflet 126

Title: Children's Future
Imagery: The leaflet depicts Afghan children engaged in education and daily activities, along with text emphasizing their bright future.
Objective: To encourage the population to support coalition forces and prevent militia activities to secure a better future for their children.

Leaflet 127

Title: Enemies No Longer in Control
Imagery: The image shows a man in flames with a large red "X" over him, symbolizing defeat.
Objective: To convey that the enemies of Afghanistan can no longer control the villages, thus demoralizing opposition forces and boosting civilian confidence in coalition efforts.

Leaflet 128

Title: You Cannot Hide
Imagery: A helicopter in pursuit, with crosshairs targeting a group of people, suggests intense surveillance and inescapability.
Objective: To send a clear message to the opposition that hiding is futile and that coalition forces have superior tracking capabilities.

Leaflet 129

Title: Coalition Forces Stability
Imagery: A family is shown with a solemn expression, highlighting the need for stability and protection.

Objective: To promote the idea that coalition forces can provide a stable environment necessary for rebuilding Afghanistan.

Leaflet 130

Title: National Duty
Imagery: The leaflet features arrows pointing from images of terrorists to images of coalition soldiers, emphasizing identification and cooperation.
Objective: To remind citizens of their national duty to inform the National Army and Coalition Forces about terrorist activities.

Leaflet 131

Title: Afghanistan's Future
Imagery: A map of Afghanistan with the national flag and a series of images showing everyday life.
Objective: To encourage the populace to take responsibility for the future of Afghanistan by supporting the government and resisting militia influences.

Leaflet 132

Title: Restoring Glory
Imagery: A scene of an Afghan city with a message overlay, evoking a sense of lost heritage.
Objective: To inspire citizens to help restore Afghanistan's former glory by supporting efforts to rebuild and reclaim the country's past.

Leaflet 133

Title: Army Provides Security
Imagery: The leaflet shows Afghan soldiers holding weapons, standing together, ready to protect.
Objective: To assure Afghan citizens that the army is there to provide security and that it is their duty to protect their family and nation.

Leaflet 134

Title: Army Helps Rebuild
Imagery: Soldiers interacting with citizens in what appears to be a supportive and collaborative effort.

Objective: To emphasize that the Afghan National Army is actively involved in the reconstruction and stabilization of Afghanistan.

Leaflet 135

Title: Afghan National Army's Commitment to Afghanistan's Future
Imagery: The leaflet displays a military convoy with soldiers in armored vehicles, representing the strength and presence of the Afghan National Army. The backdrop includes a group of children, symbolizing the future generation. The bottom panel repeats a similar scene, reinforcing the message of the Afghan National Army's role in securing and improving the future of the country. The insignia of the Afghan National Army, featuring crossed rifles and the Afghan flag, is prominently displayed, emphasizing national pride and military authority.
Objective: The leaflet aims to convey the message that the Afghan National Army is dedicated to working for a better future for all Afghans. By associating the military's presence with the well-being of children and the future, it seeks to build trust and support among the local population for the national military efforts.

Leaflet 136

Title: Distribution of Humanitarian Aid
Imagery: The leaflet shows an Afghan man picking up a leaflet from the ground with the label "HDR." On the right side, a large military aircraft is depicted dropping leaflets over a barren landscape.
Objective: The leaflet aims to inform the local population about the distribution of humanitarian daily rations (HDR) and encourage them to pick up and use the aid being provided.

Leaflet 137

Title: Halal Ration Distribution
Imagery: The leaflet shows an Afghan family sitting together, sharing a meal. They are holding HDR (Humanitarian Daily Ration) packets labeled "Halal." The background features another image of a man holding an HDR packet, smiling with the word "Halal" clearly written.
Objective: The leaflet aims to assure the Afghan people that the rations being provided by coalition forces are Halal and safe to consume according to Islamic dietary laws.

Leaflet 138

Title: Al-Qaeda, do you worry if you are safe?
Imagery: A group of armed Afghan men, presumably members of Al-Qaeda, are shown inside a cave. They

appear to be discussing something while one looks towards the sky, possibly hearing a sound. In the background, a drone is flying overhead, suggesting surveillance or an imminent strike.
Objective: The leaflet aims to instill fear and uncertainty among Al-Qaeda members by implying that they are constantly being watched and are not safe, even in their hideouts.

Leaflet 139

Title: In your graves?
Imagery: The same group of Afghan men is now in a state of panic inside the cave. An explosion has occurred, filling the cave with smoke and debris. The men are depicted as frightened, covering their heads with their hands, realizing the dire consequences of their actions.
Objective: The leaflet intends to reinforce the message of inevitable doom for those who continue to associate with Al-Qaeda, suggesting that their actions will lead them to death and destruction.

Leaflet 140

Title: Award up to 25 million dollars - Osama bin Laden
Imagery: The leaflet displays two images of Osama bin Laden, one where he is giving a speech and another showing him behind bars, implying his capture. Between the images is a stack of U.S. dollar bills, highlighting the reward amount.
Objective: The leaflet's goal is to encourage people to provide information leading to the capture of Osama bin Laden by offering a substantial financial reward.

Leaflet 141

Title: $25 Million Reward for Osama bin Laden
Imagery: The leaflet shows a text in blue background offering up to $25 million reward for reliable information about the whereabouts or capture of Osama bin Laden.
Objective: To encourage people to provide information leading to the capture of Osama bin Laden by offering a substantial financial reward.

Leaflet 142

Title: Reward for Osama bin Laden and Ayman Al-Zawahiri
Imagery: The leaflet features images of Osama bin Laden and Ayman Al-Zawahiri, each framed in red with a bold headline mentioning the reward. The text highlights the financial incentive for information leading to their

capture.
Objective: To incentivize people to provide information leading to the capture of these two key figures by offering a monetary reward.

Leaflet 143

Title: Up to $25 Million Reward for Osama bin Laden and Ayman Al-Zawahiri
Imagery: The leaflet displays a split image of Osama bin Laden and Ayman Al-Zawahiri, alongside text offering a reward of up to $25 million for information about their whereabouts or capture.
Objective: To encourage individuals to report on the whereabouts of Osama bin Laden and Ayman Al-Zawahiri by emphasizing the substantial reward.

Leaflet 144

Title: The Terrible Government of the Taliban
Imagery: The leaflet shows Haqqani, Bin Laden, and Mutawakkil with their names displayed above their heads. The background shows a dimly lit scene with the faces of these individuals highlighted, representing the Taliban government.
Objective: To discredit the Taliban government by showcasing prominent figures in a negative light, suggesting their roles in bringing terror.

Leaflet 145

Title: The Taliban Government is Falling
Imagery: The same three figures (Bin Laden, Haqqani, and Mutawakkil) are depicted, but this time their faces are altered to appear as skulls, symbolizing the fall of the Taliban regime. The text states that their rule is collapsing.
Objective: To convey the imminent downfall of the Taliban and encourage individuals to distance themselves from the group or take action against it.

Leaflet 146

Title: Osama bin Laden Laughs at You
Imagery: The leaflet shows an image of Osama bin Laden with a mocking expression, accompanied by text stating that he is laughing at those who have been sent to their deaths, implying that he is using his followers for his own gain.

Objective: To demoralize those supporting Osama bin Laden by portraying him as someone who does not value their lives and is only using them for his own agenda.

Leaflet 147

Title: Osama bin Laden sends his murderers into the world to kill for his cause.
Imagery: The leaflet features an image of Osama bin Laden with a stern expression. The text is placed next to the image, bordered with red lines on the top and bottom. The overall tone of the leaflet is serious, emphasizing the threat posed by bin Laden.
Objective: To warn individuals about the dangers posed by Osama bin Laden and his followers, encouraging vigilance and awareness of the global threat they represent.

Leaflet 148

Title: If you have information about the Taliban or Al-Qaeda, inform the coalition forces immediately.
Imagery: The leaflet shows a black and white image of a group of armed men, presumably members of the Taliban or Al-Qaeda, sitting on and around a vehicle. The image is accompanied by text in both Pashto and Dari, emphasizing the importance of providing information to the coalition forces.
Objective: To encourage local populations to report any information they have about the Taliban or Al-Qaeda to the coalition forces, thereby aiding in the efforts to combat these groups.

Leaflet 149

Title: Taliban and Al-Qaeda use innocent women and children as shields for protection.
Imagery: This leaflet depicts Taliban or Al-Qaeda fighters with women and children, emphasizing the unethical practice of using civilians as human shields.
Objective: To expose and condemn the Taliban and Al-Qaeda for their cowardly tactics, thereby turning public opinion against them.

Leaflet 150

Title: Your children are waiting for you. For peace, stability, and a bright future.
Imagery: The leaflet shows children, presumably the families of fighters, with a background of Afghanistan, appealing to the emotional bonds between fighters and their families.
Objective: The leaflet's purpose is to persuade fighters to lay down their arms and return home to their families, offering a vision of peace and a stable future.

Leaflet 151

Title: To protect yourself, do not interfere with the operations of NATO countries.
Imagery: The leaflet features a message with NATO insignia, warning against interfering with NATO operations for personal safety.
Objective: The goal is to deter locals from supporting or engaging in activities against NATO forces, emphasizing the potential danger to themselves.

Leaflet 152

Title: If you help the forces against the Taliban and Al-Qaeda and capture them, you will receive a reward of millions of dollars.
Imagery: This leaflet emphasizes the rewards available for those who provide actionable intelligence or help capture members of the Taliban or Al-Qaeda.
Objective: To incentivize collaboration with coalition forces through financial rewards, encouraging locals to act against insurgent groups.

Leaflet 153

Title: Report the presence of explosive materials to prevent attacks against minors.
Imagery: The leaflet depicts the dangers of unexploded ordnance and IEDs, encouraging people to report such materials.
Objective: The leaflet seeks to prevent injuries or deaths caused by explosives, particularly among children, by urging immediate reporting of these hazards.

Leaflet 154

Title: Mass murder all over the world.
Imagery: This leaflet lists various global terrorist attacks and their toll, including specific attacks in Nairobi, Riyadh, Bali, and Baghdad.
Objective: To highlight the global impact of terrorism and to unite people against these acts by showing the widespread devastation caused by such violence.

Leaflet 155

Title: Coalition forces are looking for Al-Qaeda members and other terrorists. If you are aware of their plans or terrorist activities, report them to these locations.
Imagery: The leaflet shows a coalition soldier interacting with locals, along with information on how to report terrorist activities safely.
Objective: To gather intelligence from local populations about terrorist activities, assuring them that their identities will be protected.

Leaflet 156

Title: Observe healthy conditions in daily life. Choose a healthy way of life.
Imagery: The leaflet provides visual instructions on purifying water by boiling it to prevent illness and death from contaminated water.
Objective: The aim is to promote public health by educating people on the importance of boiling water to eliminate germs and bacteria.

Leaflet 157

Title: You cannot hide from the water
Imagery: Image showing naval forces with a warning about placing mines in the water.
Objective: Warning against placing mines in the water and encouraging reporting of such activities.

Leaflet 158

Title: The coalition naval forces seized a ship
Imagery: Image showing a ship being seized and text warning against drug smuggling.
Objective: Discouraging support for drug smuggling and urging reporting to coalition forces.

Leaflet 159

Title: Information and news radio
Imagery: There are two leaflets with the same information. One leaflet has Pashtu and other has message in Dari language. Image of a radio tower with frequencies listed.
Objective: Informing people about radio frequencies for news and information.

Leaflet 160

Title: Support the interim government of Afghanistan
Imagery: Image of Afghan leaders with text urging support for the interim government.
Objective: Encouraging support for the interim government to ensure stability.

Leaflet 161

Title: Kashmir Khan
Imagery: Image of Kashmir Khan with smaller images depicting life in the region.
Objective: N/A (Imagery without clear objective text).

Leaflet 162

Title: Support your interim government to ensure a better life
Imagery: Image showing Afghan citizens with text encouraging support for the government.
Objective: Promoting support for the interim government to improve quality of life.

Leaflet 163

Title: Taliban: We know your location!
Imagery: Image showing a missile descending with the text warning the Taliban.
Objective: Warning the Taliban that their locations are known.

Leaflet 164

Title: Stop fighting for the Taliban and you will stay alive
Imagery: Image showing Taliban members fleeing from an attack.
Objective: Encouraging Taliban members to stop fighting to save their lives.

Leaflet 165

Title: America has provided over $170 million in aid to Afghanistan.
Imagery: The leaflet shows images of Afghan children, highlighting the positive impact of American aid in Afghanistan. On the left side, Afghan children smile and look towards the camera. On the right, a man is seen with bags of aid that are clearly labeled as provided by the USA.

Objective: The objective is to showcase the substantial aid provided by America to Afghanistan, emphasizing the significant financial assistance intended to help the Afghan people, particularly the youth.

Leaflet 166

Title: This is what the Taliban has done.
Imagery: The leaflet displays a series of images depicting the destruction caused by the Taliban. It includes images of ruined buildings, injured individuals receiving medical care, and the enforcement of harsh punishments by the Taliban.
Objective: The leaflet aims to highlight the devastation and suffering brought upon Afghanistan by the Taliban, contrasting it with the previous leaflet that emphasized U.S. aid. It serves to discredit the Taliban by showing the negative impact of their rule.

Leaflet 167

Title: Help the anti-Taliban forces to drive the murderers and thugs out of Afghanistan.
Imagery: The leaflet shows an individual, possibly representing an Afghan citizen, holding a large sum of money while pointing towards a distant mountainous area. In the background, there is an image of a group of armed fighters.
Objective: This leaflet encourages Afghan citizens to support anti-Taliban forces, promising a significant financial reward for their assistance in removing the Taliban from Afghanistan.

Leaflet 168

Title: Many tribes make a united Afghanistan.
Imagery: The leaflet features an illustration of several hands, each of a different skin tone, coming together to form the shape of Afghanistan. The visual suggests unity and cooperation among the diverse ethnic groups in the country.
Objective: This leaflet promotes national unity among Afghanistan's various tribes and ethnic groups, emphasizing that a united Afghanistan is stronger and more resilient.

Leaflet 169

Title: Afghanistan is heading towards a better future.
Imagery: The leaflet displays a map of Afghanistan with a positive message written across it, symbolizing the country's progress towards a brighter future.

Objective: The leaflet aims to instill hope and encourage support for the ongoing efforts to rebuild Afghanistan, assuring citizens that the country is on the path to improvement.

Leaflet 170

Title: Friendship: Afghanistan and the United States.
Imagery: The leaflet depicts a handshake between two hands, one representing the United States and the other Afghanistan. Above the handshake, there are images of families from both nations, symbolizing unity and friendship.
Objective: This leaflet promotes the friendly relationship between Afghanistan and the United States, emphasizing cooperation and mutual support between the two countries.

Leaflet 171

Title: The United Nations has come here to help the people of Afghanistan.
Imagery: The leaflet features the UN logo alongside a map of Afghanistan and images of Afghan citizens. The text is accompanied by an illustration of hands shaking, symbolizing international cooperation.
Objective: This leaflet aims to reassure Afghan citizens of the United Nations' commitment to aiding Afghanistan, highlighting the presence of international support in rebuilding the country.

Leaflet 172

Title: A united Afghanistan brings peace and prosperity to you.
Imagery: The leaflet shows a dove holding an olive branch, flying over the map of Afghanistan, which is shaded in the colors of the national flag. Below, there are hands clasped together in a gesture of unity.
Objective: This leaflet promotes the idea that a unified Afghanistan will lead to peace and prosperity for its citizens, encouraging national solidarity.

Leaflet 173

Title: No one has the right to determine the fate of your life.
Imagery: The leaflet features a text message stating that NATO countries are helping the people of Afghanistan to save themselves from the crimes of the Taliban and foreign terrorists.
Objective: To reassure the Afghan population that NATO's involvement is to aid in their self-determination and protect them from external threats, emphasizing the importance of their sovereignty and security.

Leaflet 174

Title: A carpet is made from several yarns!
Imagery: The leaflet features an image of a traditional Afghan carpet being woven together from multiple threads of different colors. The visual metaphor suggests that Afghanistan, like a carpet, is made stronger and more beautiful by its diverse components.
Objective: This leaflet emphasizes the importance of diversity and unity, likening the different ethnic groups and tribes in Afghanistan to the various threads that make up a strong and unified nation.

Leaflet 175

Title: Osama bin Laden and his foreign followers do not want the people of Afghanistan to live in peace with each other. Afghans must rid themselves of these fanatics.
Imagery: The leaflet shows a drawing of Osama bin Laden and his foreign followers with weapons, symbolizing threats and instability.
Objective: The message encourages Afghans to reject Osama bin Laden and foreign elements that disrupt peace and unity in Afghanistan.

Leaflet 176

Title: The time has come for Afghans to make peace among themselves.
Imagery: The leaflet depicts two Afghan men in traditional clothing, shaking hands, symbolizing peace and reconciliation.
Objective: The message promotes internal peace and urges Afghans to resolve conflicts amongst themselves, emphasizing the importance of unity.

Leaflet 177

Title: Building the Nation Together
Imagery: Two figures in traditional Afghan attire are shown building a wall together with bricks, symbolizing collaboration. The background includes mountains.
Objective: Encourage cooperation and unity among the Afghan people to rebuild the nation.

Leaflet 178

Title: A United Afghanistan Brings Peace
Imagery: A dove holding an olive branch flies over a map of Afghanistan, which is colored to represent a united nation. The image is meant to evoke peace and unity.
Objective: Promote the idea of a united Afghanistan bringing peace and prosperity.

Leaflet 179

Title: Support the New Government for New Freedoms
Imagery: A woman in traditional Afghan clothing is looking back over her shoulder. In the background, a group of men is engaged in conversation, possibly discussing government affairs.
Objective: Encourage the support of the new government by portraying it as a bringer of new freedoms for the Afghan people.

Leaflet 180

Title: Drive Out Foreign Masters and Live in Peace
Imagery: A bearded man, presumably Osama bin Laden, sits at a chessboard with skulls as pieces. The imagery suggests strategic manipulation and the high stakes of the conflict.
Objective: Urge Afghans to drive out foreign influences and live in peace, free from external manipulation.

Leaflet 181

Title: Exposing the Power Behind the Taliban
Imagery: A figure resembling Mullah Omar is shown on a leash held by Osama bin Laden, with a map of Afghanistan underneath them. This image strongly suggests the subservience of local Afghan leaders to foreign extremists.
Objective: Expose the true power dynamics within the Taliban regime, showing the influence of foreign terrorists like Bin Laden over Afghan leaders.

Leaflet 182

Title: The Commonwealth Supports Afghanistan
Imagery: A simple text-based leaflet with a blue background. The message conveys support from the Commonwealth Nations to help the people of Afghanistan.

Objective: Reassure the Afghan people of international support, specifically from the Commonwealth, in their struggle against terrorism.

Leaflet 183

Title: International Forces and Afghans Working Together
Imagery: An Afghan man in traditional attire is shaking hands with a soldier, symbolizing the collaboration between Afghan locals and international forces. The background shows a mountainous landscape.
Objective: Reinforce the message of international support, with a specific focus on the Commonwealth's role in assisting Afghanistan.

Leaflet 184

Title: "Daesh, You Are Hunted"
Imagery: ISIS fighters in crosshairs, symbolizing their targeting by Afghan forces.
Objective: To demoralize ISIS fighters by informing them that they are being actively hunted by the Afghan government, offering a chance for a better life if they surrender.

Leaflet 185

Title: End of Saddam's Oppression
Imagery: A caricature of Saddam Hussein in a military uniform holding Iraq in his hand, with a stern and oppressive expression on his face.
Objective: The message is to inform the Iraqi people that the Coalition forces are here to put an end to the oppression caused by Saddam and his regime, suggesting liberation from his rule.

Leaflet 186

Title: Tools of Suppression
Imagery: A depiction of Saddam Hussein holding Iraq with a focus on military audio fiber cords as tools of suppression, emphasizing their use to control the Iraqi people.
Objective: The leaflet warns the Iraqi people about the tools Saddam uses to suppress them, emphasizing that these are instruments of oppression under his regime.

Leaflet 187

Title: Think About Your Family
Imagery: A soldier with a distant, worried expression, surrounded by images of destruction and his family, suggesting the consequences of war on personal lives.
Objective: The message is to encourage Iraqi soldiers to think about their families and act in a way that keeps them safe, implying that continued fighting would endanger their loved ones.

Leaflet 188

Title: Before the Attack
Imagery: An illustration showing coalition aircraft preparing to strike, with a soldier contemplating the situation as artillery fire is depicted in the background.
Objective: The leaflet aims to make Iraqi soldiers think about the consequences of engaging with coalition forces, encouraging them to consider the futility of resistance.

Leaflet 189

Title: No Weapons of Mass Destruction
Imagery: A dramatic image of soldiers in gas masks against a backdrop of an explosion, with a prominent biohazard symbol, conveying the danger of using such weapons.
Objective: The leaflet warns against the use of weapons of mass destruction, highlighting the severe consequences that would follow, including retaliation from coalition forces.

Leaflet 190

Title: Accountability for WMD Use
Imagery: A stark, framed text warning against the use of weapons of mass destruction, with the implication that those responsible will be held accountable.
Objective: This leaflet delivers a stern warning to Iraqi military leaders, stating that any use of weapons of mass destruction will result in severe repercussions and that unit commanders will be held strictly accountable.

Leaflet 191

Title: People of Iraq, Please Stay Home
Imagery: A simple bordered text in Arabic providing safety instructions for Iraqi citizens.

Objective: The leaflet advises the people of Iraq to stay indoors for their safety, avoiding travel at night and not interfering with coalition operations.

Leaflet 192

Title: Success is on the Way to Liberation
Imagery: A bordered text outlining key points, including coalition gains and instructions for the Iraqi people.
Objective: This leaflet informs the Iraqi people about the progress of coalition forces, encouraging them to watch or listen to specific channels for further information.

Leaflet 193

Title: Nobody Benefits from the Use of Weapons of Mass Destruction
Imagery: A black-and-white image showing soldiers and a symbol representing radioactive danger, emphasizing the horrors of mass destruction weapons.
Objective: The leaflet warns against the use of weapons of mass destruction, highlighting the negative consequences for everyone involved.

Leaflet 194

Title: We Can See Everything
Imagery: An image of a satellite observing military activities on the ground, with text in Arabic cautioning against the use of specific weapons.
Objective: The leaflet is a warning to the Iraqi military, emphasizing that coalition forces are monitoring their activities and urging them not to use nuclear, biological, or chemical weapons.

Leaflet 195

Title: Do Not Attempt to Use Weapons of Mass Destruction
Imagery: Bordered text explaining the coalition's ability to monitor and punish those who use or transport weapons of mass destruction.
Objective: The leaflet reinforces the consequences of using or preparing to use mass destruction weapons, urging compliance to avoid severe repercussions.

Leaflet 196

Title: Do Not Interfere with Coalition Forces
Imagery: A split image showing coalition forces in action, bordered by a frame and accompanied by a cautionary message.
Objective: This leaflet discourages any interference with coalition forces, asserting that their goal is not to harm the Iraqi people.

Leaflet 197

Title: The 1st Tammuz Battalion Did Not Comply
Imagery: Two images of destroyed military vehicles, with a message highlighting the consequences of non-compliance.
Objective: The leaflet serves as a warning to other military units, showing the outcome for those who defy coalition guidelines.

Leaflet 198

Title: Do Not Release Oil into Waterways
Imagery: A bordered text warning against the environmental and economic damage caused by oil spills.
Objective: The leaflet urges the Iraqi people not to follow orders that would harm their own country, specifically warning against polluting waterways with oil.

Leaflet 199

Title: Dumping Oil Poisons Your Future
Imagery: A family standing by polluted water, emphasizing the impact of oil dumping on future generations.
Objective: To appeal to the Iraqi people's concern for their families, urging them not to contribute to environmental destruction by dumping oil into waterways.

Leaflet 200

Title: Saddam's Regime Neglected Your Family
Imagery: A concerned mother and child with a question mark above them, highlighting the uncertainty and neglect they faced.

Objective: To communicate that Saddam's regime failed to protect and care for Iraqi families, suggesting the regime's indifference to their well-being.

Leaflet 201

Title: Return to Your Home, Your Family is Waiting
Imagery: A woman holding a child, looking longingly, likely representing a mother waiting for her family to return home.
Objective: To encourage Iraqi citizens to return to their homes and reunite with their families, emphasizing the importance of family ties.

Leaflet 202

Title: Do Not Interfere for Your Safety
Imagery: Coalition forces depicted with military equipment, urging caution.
Objective: To warn Iraqi civilians not to interfere with coalition forces for their safety, indicating that the coalition is present to end Saddam's oppressive rule.

Leaflet 203

Title: Do Not Carry Weapons
Imagery: A man carrying a weapon, which could lead to misidentification as a combatant.
Objective: To advise Iraqi citizens not to carry weapons to avoid being mistaken for Saddam's soldiers or security forces, ensuring their safety.

Leaflet 204

Title: Stay Home, Do Not Carry Weapons
Imagery: A crossed-out image of a rifle, with clear instructions to stay at home and avoid carrying weapons.
Objective: To reinforce the message that the coalition forces do not wish to harm Iraqi people, encouraging them to stay home and not carry weapons for their safety.

Leaflet 205

Title: Do Not Be an Accomplice in Saddam's Crimes
Imagery: Scenes of chemical weapon victims, urging soldiers not to participate in such atrocities.

Objective: To dissuade Iraqi soldiers from using chemical weapons, warning them of personal responsibility and the consequences of becoming accomplices to Saddam's crimes.

Leaflet 206

Title: Coalition Forces Are Prepared for Chemical Attacks
Imagery: Coalition soldiers in protective gear, ready to defend against chemical attacks.
Objective: To demonstrate the coalition forces' preparedness and training against chemical weapons, discouraging any attempts to use such weapons.

Leaflet 207

Title: Coalition Forces Conducting Operations
Imagery: Coalition forces engaged in operations, with humanitarian aid being distributed.
Objective: To reassure Iraqi civilians that coalition forces are operating in their areas to provide humanitarian aid and not to harm them.

Leaflet 208

Title: Stop Repairing Military Fiber Optic Cords
Imagery: A warning framed in red, with text emphasizing the dangers of repairing military fiber optic cords.
Objective: To caution Iraqi citizens against repairing military communication infrastructure, labeling these activities as dangerous and counterproductive to their safety.

Leaflet 209

Title: "Do Not Become a Refugee!"
Imagery: Two images showing lines of displaced people, emphasizing the hardship and suffering they face.
Objective: This leaflet urges civilians to stay in their homes and villages, warning that becoming a refugee will only lead to hardship and suffering.

Leaflet 210

Title: "Coalition Forces Amass in the North"
Imagery: A large group of soldiers with helicopters in the background, suggesting a powerful military force.
Objective: The leaflet conveys the message that the coalition forces have gathered a massive fighting force in the north, aiming to intimidate and dissuade any resistance.

Leaflet 211

Title: "Facing the Coalition Means Certain Death"
Imagery: A destroyed military convoy, illustrating the consequences of facing coalition forces.
Objective: The message is clear: confronting the coalition will lead to death and the destruction of one's family. It is meant to deter any opposition to the coalition.

Leaflet 212

Title: "Warning Against Targeting Coalition Aircraft"
Imagery: Two men standing in front of a tank with leaflets raining down, with planes flying overhead.
Objective: The leaflet warns that any attempt to target coalition aircraft will result in destruction. It is a direct threat to discourage attacks on coalition forces.

Leaflet 213

Title: "Ignore This Warning, and You Will Die"
Imagery: A battlefield scene with destroyed military vehicles, emphasizing the fatal consequences of ignoring coalition warnings.
Objective: The leaflet is a stark warning that ignoring the coalition's advice will lead to death. It is meant to instill fear and compliance.

Leaflet 214

Title: "I Warned You Not to Target Coalition Planes"
Imagery: A military scene with tanks and anti-aircraft missiles, showing the aftermath of a warning ignored.
Objective: The leaflet is a follow-up, expressing regret that the warning was ignored and highlighting the consequences of such actions.

Leaflet 215

Title: "Which Fate Do You Choose?"
Imagery: A comparison of two scenarios: one where coalition aircraft destroy a target, and another where they do not.
Objective: The leaflet forces the reader to choose between compliance and destruction, emphasizing the power and reach of coalition forces.

Leaflet 216

Title: "Saddam Loses Military Support"
Imagery: Three images depicting a soldier with an Iraqi flag on his helmet, a military vehicle in action, and a desolate battlefield with soldiers walking away.
Objective: This leaflet communicates that Saddam is losing the support of his military forces. It is designed to demoralize Iraqi soldiers by showing that their leader no longer has their back, suggesting that abandoning the fight is their best option.

Leaflet 217

Title: "Saddam's Rule is Dying"
Imagery: A text-heavy leaflet bordered with a blue decorative frame, stating that "Saddam's oppressive rule is dying. It is about to end."
Objective: The leaflet aims to persuade remaining loyalists that Saddam's reign is nearing its end, urging them to abandon their positions and return to their families. It promises liberation from Saddam's brutal regime, emphasizing the inevitability of his downfall.

Leaflet 218

Title: Respond to the Coalition's Directives
Imagery: The leaflet displays a tank and soldiers in a desert landscape, with a fighter jet flying overhead. Soldiers are shown either standing or seated, suggesting a pause or a moment of decision.
Objective: To maintain your safety, respond to the coalition's directives.

Leaflet 219

Title: To Keep You Safe
Imagery: The leaflet features Arabic text bordered by a decorative pattern. There are no images, only a message emphasizing safety.
Objective: Show that you are not a threat by following coalition directions. Stay a safe distance away from anti-aircraft and artillery, but do not leave the base area. Lay down your weapons and do not interfere with coalition operations.

Leaflet 220

Title: This is How You Will Be Destroyed
Imagery: The leaflet displays two panels. The left panel shows a tank engulfed in flames, with two fighter jets attacking from above, symbolizing destruction. The right panel shows a fully intact tank with aircraft flying overhead, symbolizing the enemy's power. The imagery is framed by a bold border, with Arabic text warning about the consequences of resistance.
Objective: The leaflet warns the enemy that if they continue their aggression, they will face inevitable destruction, emphasizing the futility of resistance.

Leaflet 221

Title: "Do Not Take an Aggressive Position and You Will Not Be Destroyed"
Imagery: A tank parked safely with jets flying overhead, unharmed.
Objective: This leaflet reassures that avoiding aggressive actions will ensure safety, contrasting with the consequences shown in Leaflet 220.

Leaflet 222

Title: "Follow the Coalition's Directives"
Imagery: Military aircraft parked safely, with instructions on proper procedures, including not blocking runways and abandoning anti-aircraft equipment.
Objective: To instruct the opposing forces on how to avoid being targeted by following specific coalition instructions, emphasizing safety through compliance.

Leaflet 223

Title: "Surrender or Be Destroyed"
Imagery: Coalition ships and aircraft targeting non-surrendering Iraqi units.
Objective: The leaflet emphasizes the consequences of failing to surrender, warning that non-compliance will result in destruction by coalition forces.

Leaflet 224

Title: "The July 1st Tank Battalion's Failure"
Imagery: Destroyed tanks from the 24th Mechanized Brigade.

Objective: To illustrate the consequences of not responding to coalition instructions, using the example of the 10th Armored Division's failure.

Leaflet 225

Title: "Show You Are Not a Threat"
Imagery: Detailed instructions on parking vehicles, positioning barrels, raising white flags, and disarming to show compliance.
Objective: To provide clear guidelines for showing non-aggression to avoid being targeted by coalition forces.

Leaflet 226

Title: "Attention, Special Republican Guard Soldiers"
Imagery: A targeted image of Saddam Hussein with coalition forces coming to eliminate the regime.
Objective: To warn the Special Republican Guard soldiers not to share the fate of Saddam and his regime, urging them to abandon their allegiance.

Leaflet 227

Title: "Stay Safe: Stay Home"
Imagery: A message encouraging civilians to stay indoors and avoid interfering with coalition operations, with mention of an imposed curfew.
Objective: To instruct the civilian population to stay home for their safety and to tune in to specific radio frequencies for updates.

Leaflet 228

Title: "Coalition's Respect for Landmarks"
Imagery: Local landmarks with the message that the coalition does not want to destroy them, but will target vital military installations.
Objective: To reassure civilians that their cultural landmarks will be spared, provided they do not serve as military targets.

Leaflet 229

Title: "Targeting the Regime, Not the People"
Imagery: Military vehicles and a scene of leadership, emphasizing that the coalition's objective is the regime,

not the Iraqi people.
Objective: To make clear that coalition military operations are directed against the regime, not the civilian population, to reduce resistance from the public.

Leaflet 230

Title: "Your Families or the Regime?"
Imagery: A juxtaposition of military figures and smiling families, highlighting the choice between loyalty to the regime or the well-being of their loved ones.
Objective: To encourage soldiers to prioritize their families over allegiance to the regime, implying that the regime is not worth the sacrifice.

Leaflet 231

Title: "Luxury vs. Struggle"
Imagery: A comparison between Saddam Hussein living in luxury and the struggles of ordinary Iraqi families.
Objective: To emphasize the disparity between the luxurious life of Saddam and the hardships faced by the population, encouraging dissent among the soldiers.

Leaflet 232

Title: "Return to Your Villages"
Imagery: Families gathered together in a village setting, with a focus on children and nurturing.
Objective: To urge soldiers to abandon the battlefield and return home to take care of their families, fostering a sense of responsibility and encouraging desertion.

Leaflet 233

Title: "Don't Risk Your Life"
Imagery: A somber depiction of soldiers, with one appearing lifeless, highlighting the consequences of continuing to fight.
Objective: To dissuade soldiers from risking their lives and the lives of their comrades, promoting the idea of self-preservation over loyalty to the regime.

Leaflet 234

Title: "Think About the Future"
Imagery: A reflective soldier contemplating his future, with images of despair in the background.
Objective: To prompt soldiers to reconsider their path, questioning if the future they are fighting for is truly what they desire, and emphasizing the need to prioritize their family's well-being.

Leaflet 235

Title: "Save Yourself: Join Your Family"
Imagery: A family gathered around a meal, representing the peaceful life that awaits if the soldier chooses to abandon the fight.
Objective: To encourage soldiers to leave the battlefield and reunite with their families, emphasizing the importance of self-preservation and family unity.

Leaflet 236

Title: "Stay Away from Waterways"
Imagery: Military operations near waterways, with a focus on safety.
Objective: To advise civilians to avoid areas near waterways where military operations are taking place, prioritizing their safety.

Leaflet 237

Title: "Oil Dumps Won't Hinder Us"
Imagery: Coalition naval forces continuing operations despite oil being dumped into waterways.
Objective: To inform that sabotage efforts like oil dumping will not stop coalition operations, discouraging such tactics.

Leaflet 238

Title: "The Consequences of Oil Leaks"
Imagery: Images depicting environmental damage and economic consequences from oil leaks into waterways.
Objective: To highlight the destructive effects of oil leaks on water resources and Iraq's economic prospects, urging against such actions.

Leaflet 239

Title: "Waterways Will Be Returned"
Imagery: Coalition forces near waterways with a message of protection for civilians.
Objective: To reassure civilians that the coalition does not intend to harm them, and once Saddam's regime is defeated, the waterways will be restored to the Iraqi people.

Leaflet 240

Title: "Do Not Assist Escape"
Imagery: A ship in the water with a coalition naval vessel nearby, and a helicopter overhead.
Objective: To discourage civilians and military personnel from assisting Iraqi military leaders or regime officials who are attempting to escape, warning of severe consequences.

Leaflet 241

Title: "Attention: Do Not Help Escapees"
Imagery: A warning text box emphasizing the dangers of aiding fleeing Iraqi leaders.
Objective: To inform the public and military personnel that assisting these leaders puts them at risk and could lead to the confiscation of their boats, urging them to report defectors instead.

Leaflet 242

Title: "Your Choice"
Imagery: An explosion on the battlefield with a jet flying overhead.
Objective: To convey the stark choice between surrender and destruction, leaving the decision in the hands of the target audience.

Leaflet 243

Title: "If-Then"
Imagery: A two-part image showing a cause-and-effect scenario, where an action leads to a direct consequence (possibly retaliation).
Objective: To illustrate the immediate and inevitable consequences of resistance or aggression against coalition forces.

Leaflet 244

Title: "Military Fiber Optic Cords"
Imagery: A detailed illustration of fiber optic cords being destroyed by airstrikes.
Objective: To demonstrate the coalition's capability to target and destroy critical military infrastructure, particularly communication lines, as a warning against attempting repairs.

Leaflet 245

Title: "Repairing Puts Lives at Risk"
Imagery: Military fiber optic wires being destroyed, with a warning about the dangers of attempting repairs.
Objective: To deter individuals from trying to repair damaged military communication lines, highlighting the risks involved.

Leaflet 246

Title: "Abandon Armed Devices"
Imagery: A text-heavy leaflet instructing individuals to abandon their weapons for their safety.
Objective: To encourage the surrender or abandonment of weapons, stressing that armed devices will be targeted and destroyed regardless of their status.

Leaflet 247

Title: "Air Defense Artillery Destroyed"
Imagery: An air defense artillery unit being destroyed by airstrikes, with individuals surrendering.
Objective: To show the futility of resistance, particularly air defense efforts, and to encourage surrender to avoid further destruction.

Leaflet 248

Title: "Mining Waterways"
Imagery: A ship suspected of mining waterways being targeted.
Objective: To warn against mining waterways, informing that such actions will not impact coalition operations and will lead to the destruction of the vessel involved.

Leaflet 249

Title: "Inform Coalition of Mines"
Imagery: A ship and underwater mines with a message urging to report mining activities.
Objective: To encourage individuals to inform coalition forces about mining operations, stressing that such actions harm only the Iraqi people.

Leaflet 250

Title: "Mining the Khar Abdullah or Umm Qasr Waterways"
Imagery: The leaflet depicts a ship with potential military implications.
Objective: The message warns that mining the Khar Abdullah or Umm Qasr waterways would affect coalition ships and emphasizes that such actions would only harm the Iraqi people.

Leaflet 251

Title: "Destruction of Ships Blocking Khar Abdullah or Umm Qasr"
Imagery: The leaflet shows a ship in a waterway with a warning symbol.
Objective: It warns that any ship suspected of intending to block the passages of Khar Abdullah or Umm Qasr will be destroyed.

Leaflet 252

Title: "Coalition Forces' Respect for Iraqi People"
Imagery: The leaflet emphasizes the safety of the Iraqi people.
Objective: The coalition forces express that they do not wish to harm the proud people of Iraq and advise avoiding military presence areas to maintain safety.

Leaflet 253

Title: "Warning Against Movement from Positions"
Imagery: The leaflet discusses the risk of moving from a current position.
Objective: It warns that if individuals move from their current position, they will be attacked by coalition forces. It advises staying in position and awaiting instructions for safety.

Leaflet 254

Title: "Stay Where You Are"
Imagery: The leaflet features instructions on remaining in place.
Objective: It advises staying where you are and awaiting the coalition's instructions, warning that any movement will be considered a hostile act, potentially leading to an attack by coalition forces.

Leaflet 255

Title: "Muqteda is Nothing but a Nobody!"
Imagery: The leaflet targets a specific individual, Muqtada.
Objective: It describes Muqtada as a nobody, a coward, and a butcher, accusing him of fulfilling personal ambitions without legitimate authority. It questions his authority and motives.

Leaflet 256

Title: "Desecration of Holy Places by Muqtada"
Imagery: The leaflet discusses the desecration of holy places.
Objective: It accuses the criminal Muqtada of desecrating holy places and exploiting them as hideouts for weapons and terrorists. It calls for the return of the shrine of Imam Ali to the seminary.

Leaflet 257

Title: "Throwing Stones is Dangerous!"
Imagery: The leaflet depicts children and a warning about throwing stones.
Objective: It warns that throwing stones at coalition vehicles is dangerous and could be misunderstood as an offensive act, leading to injury or death. It emphasizes the importance of not throwing stones for everyone's safety.

Leaflet 258

Title: "Throwing Rocks is Dangerous!"
Imagery: The leaflet discusses the dangers of throwing rocks.
Objective: It warns that throwing rocks at Coalition Forces' vehicles could cause injury to soldiers or to the person throwing the rock. It emphasizes that the Coalition Forces are authorized to arrest anyone engaging in such dangerous activities.

Leaflet 259

Title: "Real Iraqi Heroes"
Imagery: The leaflet features imagery of strong, heroic figures.
Objective: It praises real Iraqi heroes, likely in a context of encouraging loyalty and patriotism.

Leaflet 260

Title: "Good Citizens of Mosul"
Imagery: The leaflet features a group of Iraqi citizens speaking with members of the Iraqi Police (IP) and the Iraqi National Guard (ING) against a blue background. The text thanks the citizens for their cooperation.
Objective: To encourage cooperation and support from the citizens of Mosul, reinforcing the security forces' commitment to making their homes safe again.

Leaflet 261

Title: "Ansar Al-Sunnah Controls Al-Qaeda"
Imagery: A graphic illustration shows a dark figure controlling a skull with a label "Al-Qaeda," symbolizing that Ansar Al-Sunnah manipulates the terrorist organization. The text describes both as takfiris.
Objective: To inform and warn the population about the extremist connections and control dynamics within terrorist groups, portraying them as mutual enemies of the people.

Leaflet 262

Title: "Al-Qaeda Controls Ansar Al-Sunnah"
Imagery: A similar design to Leaflet 261 but reverses the roles, showing Al-Qaeda controlling Ansar Al-Sunnah. The skull and dark figure imagery are repeated, emphasizing their takfiri nature.
Objective: To create confusion and distrust within terrorist networks by illustrating their manipulation of one another, portraying them as interchangeable threats.

Leaflet 263

Title: "Your Attack on Coalition Aircraft Will Lead to Your Destruction"
Imagery: An aircraft ready to launch a payload, with bombs aligned below it, and text warning against attacking coalition planes. The background features dark, ominous imagery.

Objective: To deter hostile actions against coalition aircraft by highlighting the inevitable destructive retaliation that will follow.

Leaflet 264

Title: "Do Not Shoot at the Coalition Plane"
Imagery: A leaflet with a coalition plane in the background and text warning against firing at coalition aircraft. The message emphasizes the overwhelming response from coalition forces.
Objective: To warn potential aggressors of the severe consequences of engaging coalition aircraft, giving them the choice to avoid destruction.

Leaflet 265

Title: "No Matter Where You Run"
Imagery: A close-up of eyes watching, symbolizing surveillance and the reach of coalition forces, with text asserting that the enemy will be found and brought to justice.
Objective: To intimidate and demoralize the opposition by conveying that they cannot escape the coalition's pursuit and justice.

Leaflet 266

Title: "We Have Leaders Who Are Unable to Be Good Leaders"
Imagery: The leaflet shows a somber image of leaders with a message from Abu Asim Al-Yamani, criticizing the leaders' weaknesses and poor behavior.
Objective: To sow discord and distrust within enemy ranks by criticizing their leadership and hinting at internal conflicts and failures.

Leaflet 267

Title: "We Are Coming to Destroy You!!"
Imagery: This leaflet depicts images of two men with a threatening message aimed at Al-Qaeda members, stating that they will be killed or captured.
Objective: To instill fear in the ranks of Al-Qaeda, signaling that coalition forces are actively targeting them and have insider information on their whereabouts.

Leaflet 268

Title: "Make Your City a Place of Peace"
Imagery: The leaflet features a child waving the Iraqi flag, with contact information for reporting terrorist activities, emphasizing peace and cooperation.
Objective: To encourage civilians to report terrorist activities by providing various contact channels and assuring confidentiality, promising rewards for cooperation.

Leaflet 269

Title: "The Iraqi Security Forces Are Fighting Hard"
Imagery: The leaflet displays the emblem of the Iraqi Security Forces with text thanking the citizens for their support and cooperation in the fight against terrorism.
Objective: To boost morale and support for the Iraqi security forces by recognizing the contributions of the people and encouraging continued cooperation.

Leaflet 270

Title: "Kill"
Imagery: Photos of Mohammad Jassem and Omar with a fiery background.
Objective: To urge the capture or killing of these individuals for their crimes against the Iraqi people.

Leaflet 271

Title: "Wanted"
Imagery: Multiple photos of wanted individuals with a faded background of an Iraqi flag.
Objective: To inform the public of wanted criminals and encourage providing information.

Leaflet 272

Title: "We Are All Brothers"
Imagery: Simple text on a plain background.
Objective: To promote unity and peace among Arab neighbors.

Leaflet 273

Title: "Coalition Commitment"
Imagery: Smiling Iraqi children with text above.
Objective: To emphasize the coalition's dedication to the Iraqi people and encourage support.

Leaflet 274

Title: "Liberation Support"
Imagery: Text with four circular images of prohibited acts.
Objective: To encourage support for the liberation and discourage theft and violence.

Leaflet 275

Title: "Notice for Detainees"
Imagery: Bold black text on a white background.
Objective: To inform detainees of their rights and encourage reporting mistreatment.

Leaflet 276

Title: "For Your Safety"
Imagery: Urban scene with warning text.
Objective: To advise civilians to stay indoors and avoid military areas.

Leaflet 277

Title: "Information Radio"
Imagery: Map of Iraq with radio frequencies.
Objective: To instruct citizens on how to access coalition broadcasts.

Leaflet 278

Title: "Radio for Liberation"
Imagery: Text with radio frequencies and times.
Objective: To guide citizens to tune in for important liberation updates.

Leaflet 279

Title: "Information for Liberation"
Imagery: Two boxes of text with tuning instructions.
Objective: To provide details on how to receive coalition information broadcasts.

Leaflet 280

Title: "Information Television"
Imagery: Map of Iraq with a marked area and instructions about tuning television antennas.
Objective: Inform Baghdad families to adjust their television antennas to receive important information from Coalition Forces on Channel 3.

Leaflet 281

Title: "Information Radio"
Imagery: A map of Iraq with a radio tower, frequencies, and times displayed.
Objective: Guide the Iraqi population on how to tune into the coalition's information radio broadcast for important updates.

Leaflet 282

Title: "Stay at Home"
Imagery: A family sitting together in their home, suggesting a peaceful environment.
Objective: Encourage Iraqi citizens to stay at home with their families to ensure safety and peace during the ongoing conflict.

Leaflet 283

Title: "Stay Safe"
Imagery: Text within a bordered frame, providing detailed instructions.
Objective: Advise the people of Iraq to stay home and avoid outdoor activities, especially at night, to avoid being mistaken for military targets by coalition forces.

Leaflet 284

Title: "Civilian Protection Instructions"
Imagery: A figure demonstrating how to protect oneself when hearing an approaching plane.
Objective: The leaflet provides specific instructions for civilians to protect themselves during airstrikes, emphasizing that coalition forces do not target civilians.

Leaflet 285

Title: "Da'esh Destruction"
Imagery: Rubble with a Da'esh flag amidst the ruins, highlighting the destruction caused by ISIS.
Objective: The leaflet advises civilians to keep away from Da'esh buildings, as coalition forces are targeting them for destruction.

Leaflet 286

Title: "ISIS Steals Childhood"
Imagery: A split image showing a young boy being forced to execute someone and a group of children playing, contrasting innocence with corruption.
Objective: The leaflet condemns ISIS for turning children into murderers, urging the protection of their innocence.

Leaflet 287

Title: "Coalition Pilots Prowl the Skies"
Imagery: A coalition pilot with an explosion in the background, symbolizing the airstrikes against ISIS.
Objective: The leaflet warns ISIS fighters that coalition pilots are targeting them, encouraging civilians to recognize the threat posed by ISIS.

Leaflet 288

Title: "Daesh, You Are Hunted"
Imagery: A caricature of a Daesh fighter fleeing from a swarm of angry bees emerging from a beehive painted with the Iraqi flag colors.
Objective: To mock Daesh and portray them as cowardly and easily defeated.

Leaflet 289

Title: "Fallujah"
Imagery: A lion representing Iraq pouncing on a snake labeled ISIS, with the word "Fallujah" written in the background.
Objective: To symbolize the strength and victory of Iraqi forces over ISIS in Fallujah.

Leaflet 290

Title: "Iraqi Security Forces Advancing"
Imagery: Iraqi soldiers advancing with determination, with text suggesting rapid progress, similar to their efforts in Fallujah.
Objective: To convey the momentum and strength of Iraqi forces in their military operations.

Leaflet 291

Title: "Operation Inherent Resolve"
Imagery: A vehicle with armed fighters being hit by drones, with the fighters being thrown out of the vehicle.
Objective: To depict the effectiveness of the Operation Inherent Resolve campaign in targeting and eliminating ISIS militants.

Leaflet 292

Title: "For Any Oppressor, There is an End"
Imagery: A Kurdish fighter shooting an ISIS militant, symbolizing the defeat of oppression.
Objective: To emphasize the inevitability of defeat for oppressors and to bolster the morale of anti-ISIS forces.

Leaflet 293

Title: "End of Daesh"
Imagery: A dark figure representing Daesh, surrounded by the remains of fallen soldiers.
Objective: To symbolize the eventual demise of Daesh and its fighters.

Leaflet 294

Title: "ISIS Uses Human Shields"
Imagery: A depiction of ISIS militants using civilians as human shields to protect themselves from airstrikes.
Objective: To expose and condemn the tactics of ISIS, highlighting their disregard for human life.

Leaflet 295

Title: "Attention Syrian People"
Imagery: Coalition aircraft targeting ISIS fighters while civilians are shown fleeing the area.
Objective: To warn civilians to distance themselves from ISIS militants to avoid being caught in airstrikes.

Leaflet 296

Title: "Leave the Fight Now"
Imagery: ISIS fighters fleeing from a vehicle, dropping their weapons and banners.
Objective: To encourage ISIS members to abandon the fight and save themselves from inevitable defeat.

Leaflet 297

Title: "Don't Let Daesh Take You"
Imagery: Civilians being dragged by ISIS fighters, while others resist and try to escape.
Objective: To discourage people from joining or staying with ISIS, suggesting they will meet a grim fate if they do.

Leaflet 298

Title: "The Time to Leave Raqqa"
Imagery: Civilians fleeing a war-torn area with a banner of ISIS in the background, symbolizing the urgency to escape.
Objective: To encourage civilians in Raqqa to leave the city, warning them that the time to flee has arrived.

Leaflet 299

Title: "Reserve Your Place Before It's Too Late"
Imagery: A bus with an outstretched hand reaching toward it, symbolizing an offer of safety to those who

surrender.
Objective: To persuade individuals involved in carrying weapons to surrender before it's too late and reserve their place on the "road to salvation."

Leaflet 300

Title: "Waving the White Flag"
Imagery: Soldiers of the Syrian Democratic Forces observing a civilian waving a white flag, indicating surrender.
Objective: To instruct those who do not wish to fight to surrender to the Syrian Democratic Forces, promising fair treatment.

Leaflet 301

Title: "Airstrikes on ISIS Transportation"
Imagery: Images of boats and ferries with red prohibition symbols, indicating they are targets for airstrikes.
Objective: To warn against using boats or ferries for transportation, as they are being targeted by airstrikes.

Leaflet 302

Title: "Departure Instructions"
Imagery: A written instruction card with detailed steps for safe departure.
Objective: To provide detailed instructions for safe surrender and departure to those fleeing Raqqa, emphasizing secrecy and safety.

Leaflet 303

Title: "Daesh Will Lose Tabqa"
Imagery: A burning building with the names of cities that have already fallen, including Raqqa and Manbij.
Objective: To declare that ISIS will lose control of Tabqa, just as it did in other cities, urging the residents to take action.

Leaflet 304

Title: "Removing Unexploded Materials"
Imagery: A soldier in protective gear handling unexploded ordnance, with text about the civil council's efforts.

Objective: To inform the residents of Tabqa and Raqqa about efforts to remove unexploded materials and how they can contribute by reporting locations.

Leaflet 305

Title: "Stay Away from Daesh Locations"
Imagery: A map with highlighted danger zones and images of ISIS fighters in these areas.
Objective: To warn civilians to avoid specific locations used by ISIS, as they are targets for coalition strikes.

Leaflet 306

Title: "Warning: Stay Away"
Imagery: A military convoy targeted by an airstrike, marked with a red crosshair.
Objective: Warn civilians to avoid military activities and stay clear of areas where such activities are taking place to ensure their safety.

Leaflet 307

Title: "NATO Protects Civilians"
Imagery: A NATO shield surrounded by civilians and soldiers, symbolizing protection.
Objective: Convey that NATO forces are in the region to protect civilians and maintain peace.

Leaflet 308

Title: "Help Has Arrived"
Imagery: A crowd of people waving flags and celebrating, with the text "Help Has Arrived!" displayed prominently.
Objective: The leaflet aims to encourage the population to spread the word that coalition forces are conducting military operations to bring peace.

Leaflet 309

Title: "Hope for Benghazi"
Imagery: Text against a dark backdrop, urging the people of Benghazi to take hope as international forces are working to stop the violence.
Objective: To reassure the people of Benghazi that the coalition is enforcing a "No Fly Zone" to protect civilians and bring peace.

Leaflet 310

Title: "NATO Air Forces Enforce No-Fly Zone"
Imagery: An unmanned aerial vehicle (UAV) targeting a location in Libya with NATO branding.
Objective: To inform the public that NATO air forces have imposed a no-fly zone to protect Libyan civilians.

Leaflet 311

Title: "One Libya, One People"
Imagery: A composite image showing a soldier and a civilian unified as one, with the Libyan flag in the background.
Objective: To promote national unity in Libya by urging citizens to stop tearing the country apart and work together as one people.

Leaflet 312

Title: "Soldiers Committing Crimes Against Humanity"
Imagery: A group of soldiers with an ominous statue of Gaddafi in the background, emphasizing the message of accountability.
Objective: The leaflet warns soldiers that those committing crimes against humanity will be held accountable by the international community, urging them to choose peace and prosperity over continued war.

Leaflet 313

Title: "Libya is One"
Imagery: A radio tower with waves emanating from it, signifying unity and communication across the nation.
Objective: The leaflet encourages Libyans to tune into a specific FM station, emphasizing the message that Libya is one and its people are united.

Leaflet 314

Title: "Freedom of Information"
Imagery: A radio tower broadcasting signals, highlighting the importance of information access.
Objective: The leaflet calls on Libyans to demand their legal right to free information, directing them to tune in to a specific FM station.

Leaflet 315

Title: "Warning to Gaddafi's Leaders"
Imagery: A map with a red circle and a warning sign, indicating a targeted area.
Objective: The leaflet warns the leaders of Gaddafi's regime in Sirte to stop fighting and leave immediately to avoid further conflict.

Leaflet 316

Title: "Reconciliation and Peace"
Imagery: A puzzle piece of Libya's flag coming together, symbolizing the potential for unity.
Objective: The leaflet promotes reconciliation, justice, freedom, and peace as the foundations for Libya's future.

Leaflet 317

Title: "Libya is One and its People are One"
Imagery: Two men embracing under a dove of peace, set against a backdrop of Libyan landmarks.
Objective: The leaflet reinforces the message of national unity, urging Libyans to see themselves as one people.